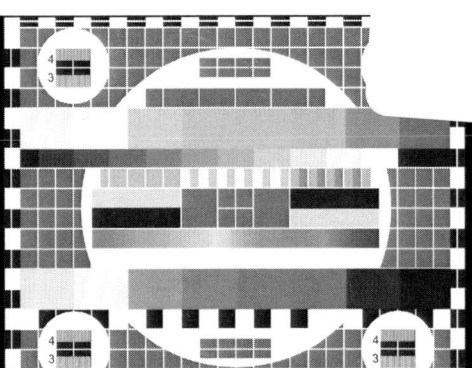

TELEVISION TECHNICAL THEORY

Dana M. Lee
Ryerson University

Cover images © Shutterstock, Inc.

Kendall Hunt
publishing company
www.kendallhunt.com
Send all inquiries to:
4050 Westmark Drive
Dubuque, IA 52004-1840

Copyright © 2010 Kendall Hunt Publishing Company

ISBN 978-0-7575-7319-4

All rights reserved. No part of this publication may be reproduced, stored in a retrieval system, or transmitted, in any form or by any means, electronic, mechanical, photocopying, recording, or otherwise, without the prior written permission of the copyright owner.

Printed in the United States of America
10 9 8 7 6 5 4 3 2 1

Contents

1. WHAT'S IN A TELEVISION STATION   1
2. TELEVISION: A BRIEF TECHNICAL HISTORY AND OVERVIEW   7
3. ELECTRICITY 101   15
4. HOW IT ALL CONNECTS TOGETHER   19
5. ANALOG VIDEO   33
6. DIGITAL VIDEO   49
7. MEASURING VIDEO   73
8. MONITORS AND TELEVISION SETS   93
9. CAMERAS   111
10. LIGHTING   137
11. SPECIAL EFFECTS   155
12. VIDEO RECORDING AND REPRODUCING   173
13. EDITING   189
14. FILM FOR TELEVISION   195
15. TRANSMISSION   199
16. SECTION REFLECTIONS   223

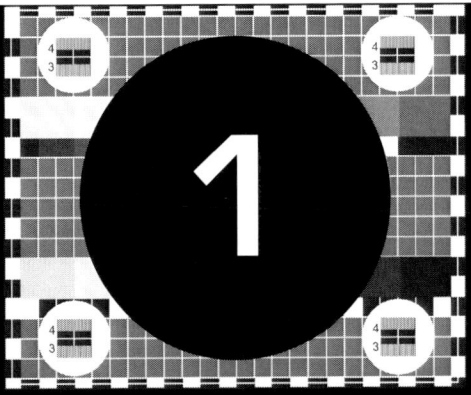

WHAT'S IN A TELEVISION STATION?

"Photography is going to marry Miss Wireless, and heaven help everybody when they get married. Life will be very complicated."

—Marcus Adams, Society photographer, in the *Observer*, 1925

2 ■ Television Technical Theory

THE TV STATION AS A WHOLE

The basic television system consists of equipment, and people to operate this gear so that we can produce TV programs. The stuff you'll find in a television station consists of (and this list is not exhaustive!):

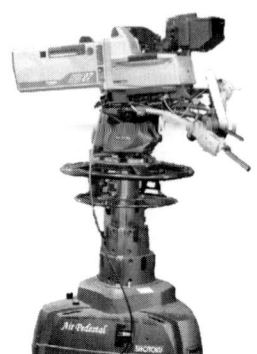

◀ One or more television cameras

Lighting, to see what we're shooting ▶

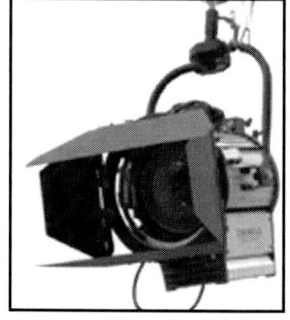

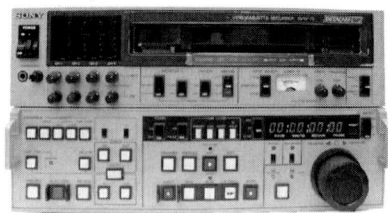

◀ One or more video recorders, in any of a number of formats

One or more video switchers, to select video sources, ▶
to perform basic transitions between those sources,
and to create special effects

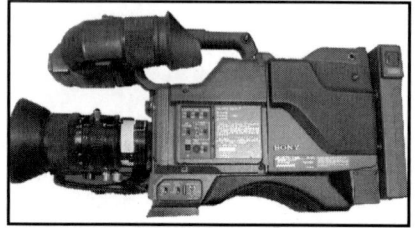

◀ EFP (electronic field production) shooting and
production equipment and storage facilities

◀ A postproduction editing facility, to assemble program segments together

Some special effects, either visual or aural—electronic, optical, or mechanical

One or more audio consoles, along with sound ▶ control equipment, to manipulate the sounds we generate with microphones, audio recorders and players, and other devices

THE STUDIO INFRASTRUCTURE

Whether or not you work in a traditional studio or in a "studioless" environment (for example, an on-air newsroom), the same principles apply. You'll still need:

◀ An intercom system (with headset stations for floor directors)

Floor monitors (video and audio) ▶

Electrical outlets both for technical equipment and lighting and for equipment that runs on regular AC power

In addition, your control room or control centre will have:

◀ Various program and preview monitors

Program audio speakers

4 ■ Television Technical Theory

◄ Time of day clock

An audio control room, with audio ► console, CD player and/or other digital recording and playback technology, and auxiliary audio enhancement equipment

OTHER AREAS

You will also need the services of:

- Central VTR, with its various pieces of video recording equipment
- Master control, without which the television program can't get on the air
- An engineering department, to keep all of this running and to design and build new and improved facilities

Master control

In this textbook we will be exploring all of this equipment and all of these areas.

Chapter 1 What's In a Television Station? 5

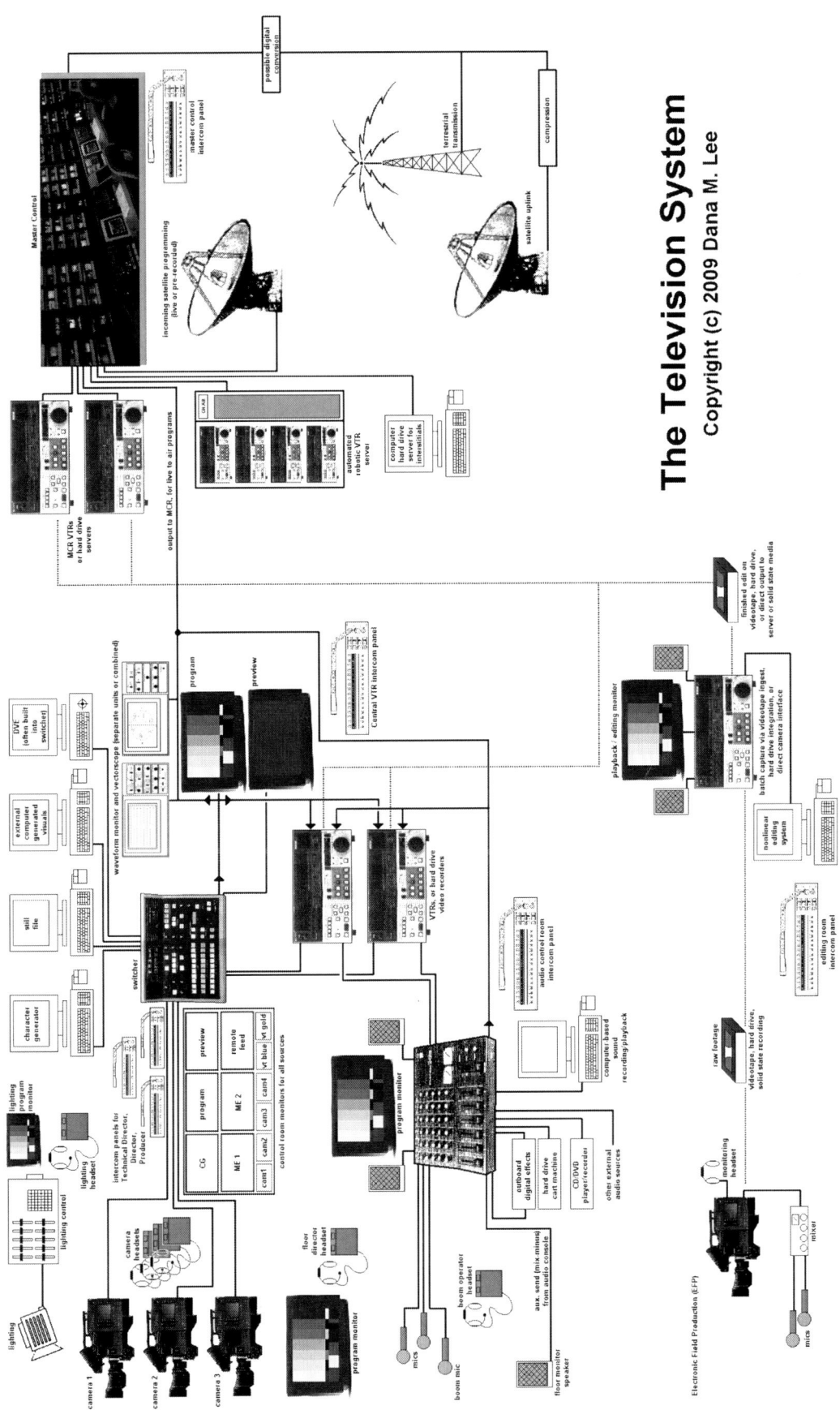

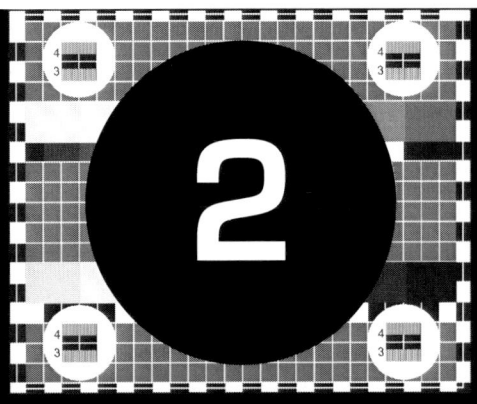

2

TELEVISION
A BRIEF TECHNICAL HISTORY AND OVERVIEW

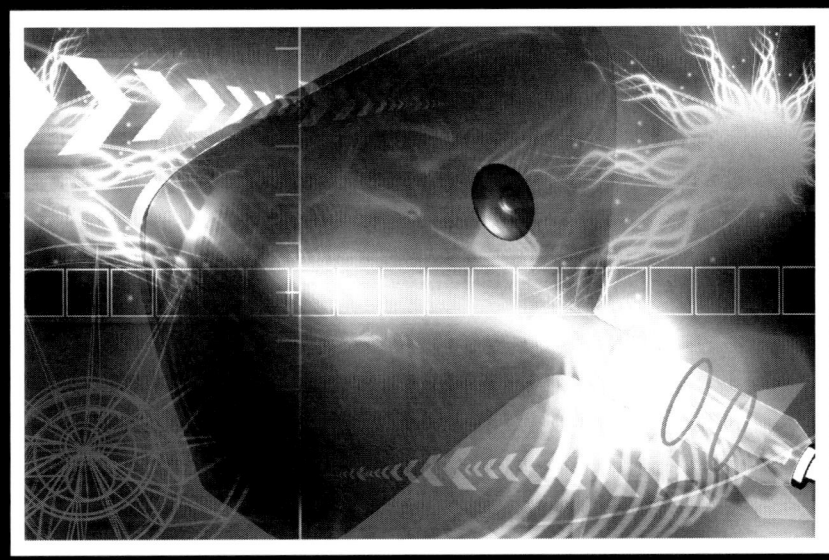

"If it weren't for Philo T. Farnsworth, the inventor of television, we'd still be eating frozen radio dinners."

—Johnny Carson

TELEVISION'S TRAINING GROUND

A View to the Past

Television. Have you ever stopped to consider what impact our industry has? Even though it's only been fully established for about sixty years, the advances in it are astounding. Fifty years ago the images were small, blurry, and black-and-white, with one channel of audio. Now, they're big, bright, in colour, and in high definition (HD). The sound can be stereo audio or 5.1 surround sound. We take cable TV, home satellite dishes, DVDs, PVRs, and mobile TV for granted.

Video has become a dominant force in the lives of Canadians:

- 99% of Canadians watch television, and the average viewer watches approximately 25 hours a week or on average more than 3 hours every day[1]
- 99% have one or more television sets[2]
- 20% of those own an HDTV or an HDTV receiver[3]
- 79% of Canadian viewers have a VCR[2]
- 69% have cable TV[2]
- 24% have a satellite dish[2]
- 85% own a DVD player[2]
- 13% own a PVR[4]

Canadians make their own videos from cameras and camcorders, buying millions of blank DVDs per year. They also store up TV programs for later viewing, recording them on their PVRs. They buy or rent movies or other programs on DVD—again, millions of times a year. Even older media such as motion pictures increasingly depend on video technology for production and distribution.

Since its start in the 1920s, broadcasting in general has provoked a variety of responses from institutions affected by it. Businesses often see it as an important contemporary advertising medium, but many educators still regard it as a foe of literacy and of serious thought. Critics have pointed out its tendency toward escapism. All of this is debatable, but without a doubt, television has become part of our lives. It brings the world's events to us whenever and almost wherever we want. It has documented historical moments in history, such as man's first steps

---

[1] BBM, Nielsen Media Research, Statistics Canada: 2007.

[2] "Spending Patterns in Canada," Statistics Canada: 2007.

[3] CBC/Radio-Canada's MTM: 2005.

[4] "The Battle for the North American (Canada & US) Couch Potato," Convergence Consulting Group: 2009.

on the moon, and has brought us colour pictures from planets in the solar system. Whatever its achievements, whatever its failings, there's no doubt that television, and video productions in general, are capable of exerting an influence that almost no other medium of communication can match. This is something to think about the next time you pull a dissolve lever, play a video, aim a camera, slide an audio fader, count out an interstitial segment, or make an edit.

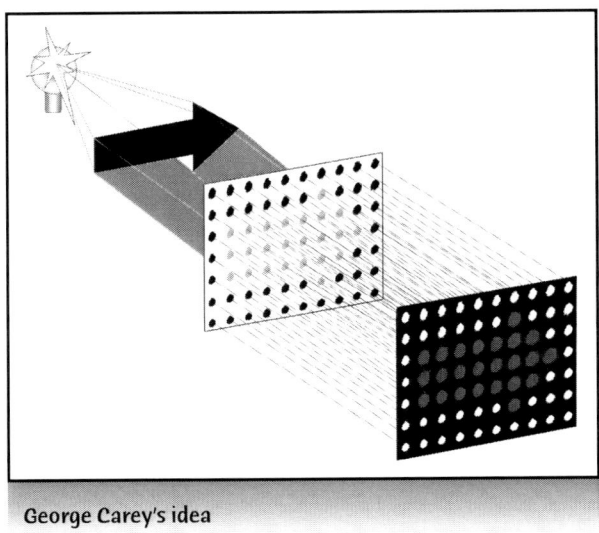

George Carey's idea

How It All Began

The basic concept of television—the transmission of images over distances—had challenged scientists even before the invention of movies or radio.

In 1875 in Boston, George Carey proposed a system consisting of a grid of photoelectric cells facing the image to be transmitted, connected in parallel to a similar grid made up of light bulbs. The picture to be transmitted would form a pattern of electrical charges on the photocell matrix, which in turn would be sent to the light bulbs, recreating the original image. It was, by definition, a very low resolution system, and would have required thousands of connections from the photocells to the bulbs.

The Nipkow scanning disc, the basis of mechanical television, was designed in 1883. The plan included a simple metal disc perforated by holes arranged in a spiral. When it revolved, the disc would scan a light beam on an object or picture placed behind it. The light would reflect off brighter parts of the image, and be absorbed by the darker areas. The resulting changes in light intensity would be picked up by a photoelectric cell, which would convert the changes in light into an electrical signal. This signal would be sent via electric wires to a receiver, where there was an identical disc turning at the same speed in front of a lamp whose brightness changed according to the received signal. The result of this rapidly changing picture would have been the illusion of movement. Paul Nipkow's ideas were just that—ideas—since this process couldn't actually be implemented with the technology available in the late nineteenth century.

The present system of electronic television was proposed in detail by a Scotsman, A. A. Campbell-Swinton, in 1908. But his ideas also were theoretical, since the ability to manufacture cathode ray tubes wasn't yet ready for prime time.

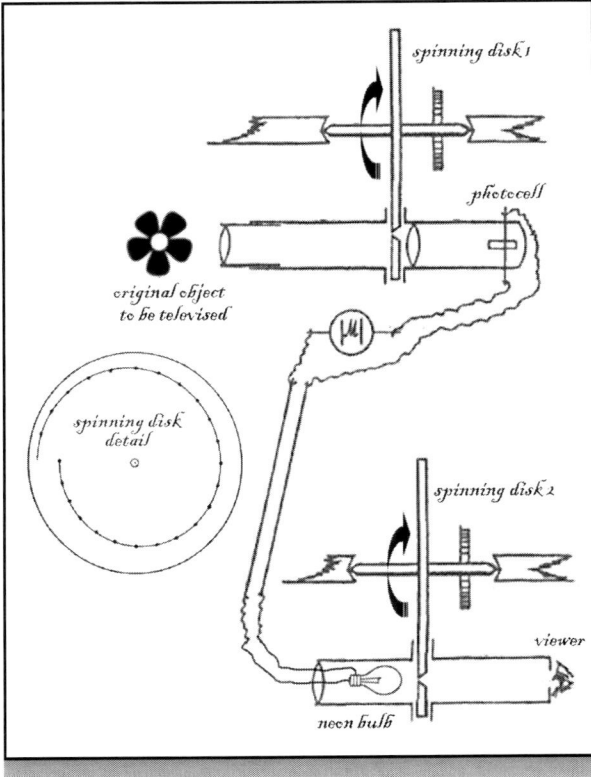

The Nipkow disc system (after his original patent drawing)

Baird's mechanical system Televisor

As an aside, the many early television schemes envisioned transmission through wires, not over the air. Guglielmo Marconi's invention of the wireless radio (1895) spurred efforts toward over-the-air transmission of pictures, and by the late 1920s, radio and motion pictures were combined. In 1928 Charles Francis Jenkins, the inventor of the modern motion picture projector, began regular broadcasts of crude "radiomovies" in Washington, D.C., using motion-picture film as a source.

John Logie Baird had developed a similar mechanical TV system in Britain in 1926, based on Nipkow's concepts, which became the basis for the BBC's first regular television broadcasts. Also in the 1920s, Baird developed and demonstrated the first colour television system and the first videodisc system by converting the video signal into audio and then recording it on phonograph records.

The End of Spinning Disks

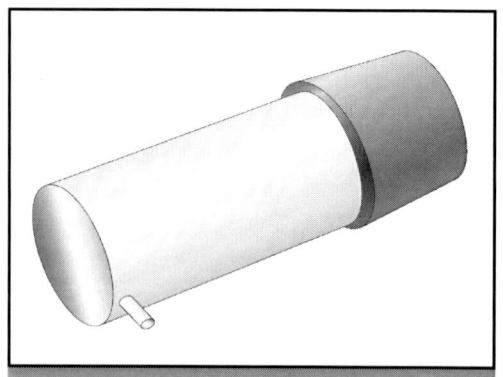
Farnsworth's image dissector tube

Electronic television, which used cathode ray tubes as receivers and transmitters, was developed simultaneously and independently in the United States in the early 1920s by Vladimir K. Zworykin and Philo T. Farnsworth. They both built on the electron tube theories developed in 1897 by Karl Ferdinand Braun in Germany.

The first regular broadcasts of electronic television for the public began in 1936, in London, England, using 405 horizontal scanning lines. France adopted a 455-line electronic system that year. Also in 1936, the Berlin Olympics were telecast with 441 lines, and watched in special viewing rooms by the people of Berlin and Leipzig.

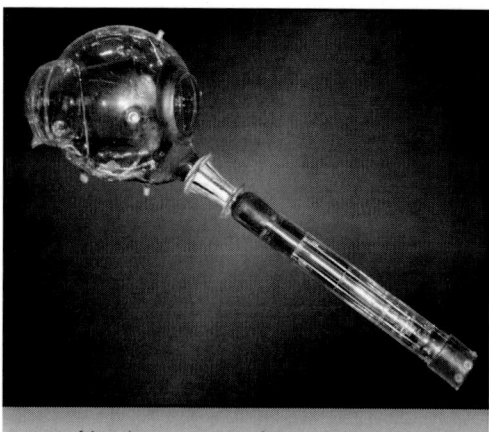
Zworykin's iconoscope tube

In the United States, the first public showing of television was at the 1939 World's Fair in New York, using a 340-line system. Two years later, in 1941, an industry-wide engineering committee in the U.S. adopted standards for a 525-line system based on specifications developed by the Radio Corporation of America (RCA). The NTSC system (named after the National Television System Committee, which developed it) was authorized by the Federal Communications Commission (FCC). At that time there were about 7,000 television sets in the U.S., most of them in New York City. However, during the Second World War, almost all electronic technology was diverted to the war effort, resulting in broadcasting being reduced from fifteen hours a week to just four.

It was after World War II, in 1945, that regular broadcasting began in earnest and television developed rapidly throughout the world. Most of Europe, though, chose a 625-line system that was incompatible with the U.S. 525-line standard. When peacetime activities resumed, television mania quickly swept the continent, on both sides of the North American border.

The Launch of Colour Television

Astoundingly, and over the objections of much of the U.S. television industry, the FCC in 1950 approved a mechanical colour television system that was incompatible with the millions of black-and-white sets then in use. It had been developed by the Columbia Broadcasting System (CBS) and involved placing rotating coloured filter wheels over both the camera's imaging device and the home receiver. Eventually, a second National Television System Committee was convened to develop a black-and-white-compatible colour system. The 525-line NTSC colour system gained FCC approval in 1953, but it was ten years before the public responded and bought colour sets in any significant numbers. In Europe, two different 625-line colour systems were introduced, PAL and SECAM.

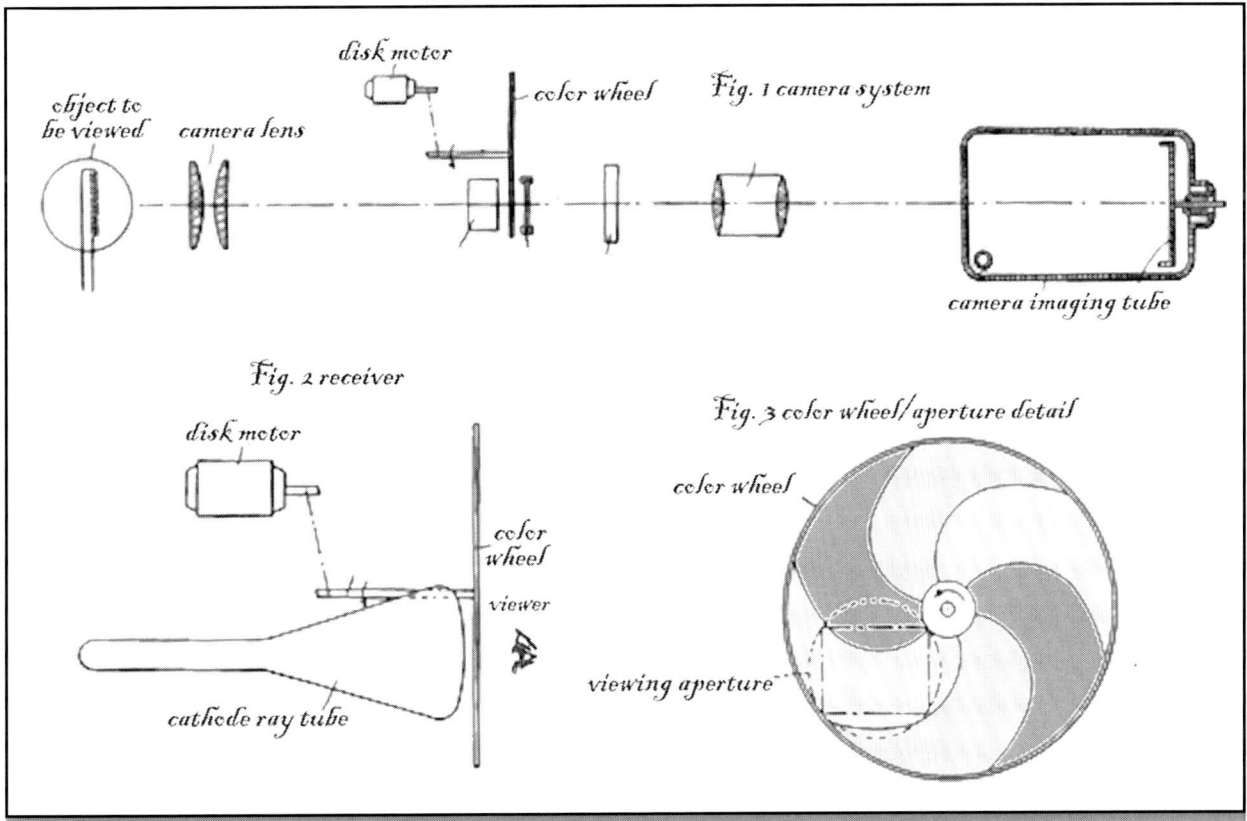

CBS' proposed (and, for a short while, approved) mechanical colour television system (after CBS' patent drawing)

Videotape Production

Prior to 1956, there were no videotape recorders. All recordings were made with a 16mm film camera shooting the images on a television monitor, using a system called a kinescope. The quality of these recordings was, at best, mediocre. Ampex Corporation exhibited the first videotape machine at the CBS Television Affiliates Meeting in Chicago on April 14, 1956. It was first used on-air on November 30, 1956, to time-delay "Doug Edwards and the News" for CBS affiliates on the West Coast of the U.S.

At first, videotape editing involved carefully splicing the two-inch-wide magnetic tape with a razor blade and special adhesive tape. The world's first electronic videotape editing system was invented in 1962 by a Canadian postproduction house, Advertel Productions. Two years later, this same editing system was marketed by Ampex under the name "Editec."

The Launch of Digital Television

As far back as the late 1980s, engineers were working on the possibilities of fitting a digital television signal into the existing analog television channel space. The object of this research was to comprehend the possibilities of transmitting a high-quality picture and multichannel sound to the home. After several years of research and development, for not only digital television but high definition video as well, approval was granted by the FCC. The first HDTV transmission in North America occurred on WRAL channel 32 in Raleigh, North Carolina, on July 23, 1996.

So, What about Canada?

In Canada, the first television station was VE9EC in Montreal, jointly owned by radio station CKAC and the newspaper La Presse. It went on the air on October 9, 1931, with neon-red pictures. One small problem—there were practically no receivers available to view the flickering images. Canadian Television Limited eventually collapsed during the Depression.

During the 1940s, the Marconi Company in Montreal and Famous Players Theatre in Toronto applied for television licences. CFRB in Toronto had applied in 1938 and was denied. CKEY followed, with the same fate, in 1945. They were turned down by none other than the Canadian Broadcasting Corporation (CBC), which ran the government-owned public broadcasting system and was also the government licensing organization, deciding who the competition would be.

CBC owned 25 radio stations coast to coast—the Trans-Canada Network—which also included another 30 privately owned stations. This was distinct from the Dominion Network, which linked almost 50 private radio stations that ran commercial material, including several American radio shows. And then there were the independents, such as CFRB, and those affiliated with American networks. This was the foundation upon which Canada's television industry would be built.

After the war, television in North America was slow in starting back up. But by 1951, there were twelve million television receivers in the U.S. and five networks: NBC, CBS, ABC, DuMont, and Westinghouse. American television continued to attract Canadian audiences. Many border television stations sprang up to beam their signals into parts of Ontario. WBEN Buffalo, WROC Rochester, WHEN and WSYR Syracuse, WWJ, WXYZ, and WJBK Detroit, all these signed on between 1947 and 1950. Vancouver, meanwhile, was invaded by KING Seattle in 1948. By 1951, there were more than 90,000 television sets in Canada—all watching American stations.

Finally, on September 6, 1952, the CBC went on the air with CBFT channel 2 in Montreal, with a newsreel, a bilingual variety show, a documentary about Montreal, and a ceremonial inaugural program. The evening finished with a film about television ("Kaleidoscope") and a French production of "Oedipus Rex."

Two days later, on September 8, CBC Toronto signed on with CBLT channel 9. The first televised image was the station ID slide—shown upside down and backward. The first film to be shown broke in the projector's gate. The programming evening began with "Let's See" hosted by Percy Saltzman. Following this was the news (featuring a news story involving the famous "Boyd Gang" which had broken out of the Don Jail and had been recaptured). The official Toronto opening was celebrated, and finally the under-three-hour broadcast day ended with a variety program from Montreal.

In 1952, there were an estimated 146,000 television sets being watched in Canada. This grew rapidly and, by 1955, there were 1.4 million television receivers in the country. Not all of them, however, were watching the CBC. There were 26 English television stations on the air (six owned by the CBC), and four French stations (two owned by the CBC.)

CBC television was established to produce television programs, sell television receivers, and build a large and loyal audience. By 1955, more than 70 percent of all Canadians were able to receive one of CBC's stations. They had built the audience, but couldn't convert many of the Canadian-based viewers of U.S. programming. Interestingly, while the Canadians were complaining about the CBC production quality, U.S. television salesmen were selling TV sets south of the border by telling viewers how good Canadian television programs were!

While the FCC approved the NTSC colour television standard in 1953, it was not until 1967 that colour production was an accepted standard in Canada.

Other Networks and Stations

In January 1960 the Board of Broadcast Governors (BBG, precursor to the Canadian Radio-television and Telecommunications Commission or CRTC) began hearings for a series of television stations that would be allowed to broadcast in eight major Canadian markets—Halifax, Montreal, Ottawa, Toronto, Winnipeg, Calgary, Edmonton, and Vancouver. The stations that received these licences, under the guidance of Spencer Caldwell, eventually formed the Canadian Television Network (CTN), renamed CTV in 1962.

Although CHCH Hamilton went on the air in 1954 as a CBC affiliate, it severed its link to the CBC on October 1, 1961, and became Canada's first independent television station. Over the years the station has produced many local television programs for its audience.

On September 28, 1972, an independent station of a different sort emerged—Citytv, channel 79 in Toronto. Licensed to broadcast uniquely different and alternative local programming, City was known for its self-proclaimed "gutsy television," featuring many Toronto-oriented programs and late-evening soft-core pornography.

Also in 1972, Global Television, an Ontario regional network, was licensed. It went live on January 6, 1974, and originally consisted of six transmitters in Ottawa, Bancroft, Uxbridge, Sarnia, Paris, and Windsor. By the late 1980s, more Ontario transmitters, as well as other stations in Canada, joined the group, and in 1997 Global branded itself as a national network.

In 1982, the first pay-TV specialty channels were approved by the CRTC: First Choice (a movie network), Superchannel (general interest programming), C-Channel (Canadian culture), and a multilingual service in British Columbia. In 1984, additional specialty services were launched: MuchMusic, TSN, TeleLatino, Cathay, and Chinavision.

The Present... and Future

Today, Canadian television viewing is fragmented between specialty stations (45%), CTV (15%), other Canadian and U.S. conventional broadcasters (12% each), and CBC and Canwest (both with 8% each.) Most Canadians —93%—view television either through cable or satellite.[5]

Video watching in general, of course, has come a long way from simply turning on the television to receive a terrestrial broadcast with an antenna. Today, video is distributed by over-the-air transmissions, cable, satellite, wireless cable, IPTV, and the Internet. The channels of distribution include conventional television stations, specialty TV, pay-per-view, video on demand (VOD), DVDs, PVR recordings, as well as video downloads and streams over the Internet. Today's distribution systems and channels are no longer independent—each of these can be easily altered from one format to another.

And finally, the simple television set itself is no longer what it once was. It has morphed into a video monitor connected to cable, satellite, a VCR, or a PVR. In fact, that screen doesn't have to be a television set at all—other viewing options include digital media players, mobile phones, or laptop computers.

---

[5] Nielsen Media Research: 2007.

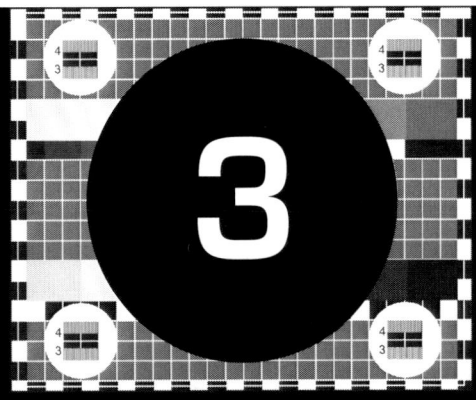

ELECTRICITY 101

"I introduced into my ears two metal rods with rounded ends and joined them to the terminals of the apparatus. At the moment the circuit was completed, I received a shock in the head and heard a crackling and boiling noise. This disagreeable sensation, which I feared might be dangerous, has deterred me so that I have not repeated this experiment."

—Alessandro Volta

Electricity and the Television System

The full electronic television system consists of a sometimes incomprehensible jumble of wires and connections. Sometimes these connections fail, often for very simple reasons. Without an understanding of how basic electrical circuits work, it's difficult to see how our television system fits together, or how we might solve simple problems in the studio or on location. So, this chapter (along with "How It All Connects Together") will introduce some of these fundamental concepts.

A basic electric circuit is a series of electrical components connected together, forming a path for electricity to flow. The purpose of most circuits is to convert electricity into a different form of energy, such as light, heat, or sound. All electric circuits have four main parts:

- A source of electric energy, such as a chemical battery, generator, solar cell, or an AC outlet in the wall
- A load (also known as an output device), such as a lamp, motor, loudspeaker, or video monitor
- Conductors, such as copper or aluminum wire, to transport the electrical energy from the source to the load
- A control device, such as a switch, relay, or control knob, to control the flow of energy to the load

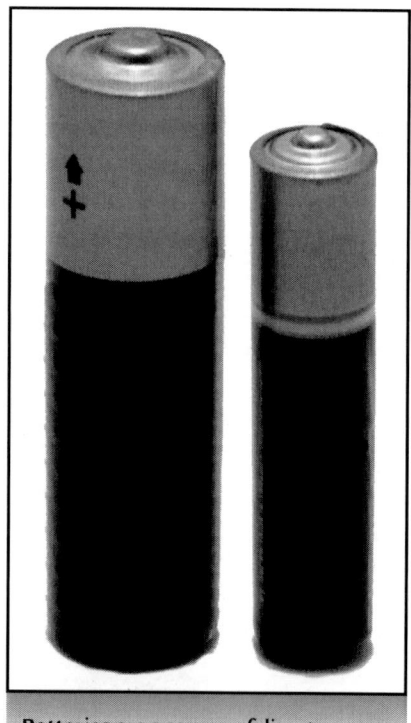

Batteries are a source of direct current

A household outlet is a source of alternating current

The source may be either DC (direct current, which does not vary in value) or AC (alternating current, which periodically reverses its polarity). Direct current is found in batteries or solar cells; alternating current is what is available from regular household AC outlets or from an on-location generator. In North America, alternating current reverses its polarity at a rate of 60 Hz (cycles per second); in the UK and other countries, it fluctuates at 50 Hz.

Series Circuits

A series circuit is one in which the current has only one path to take—from one side of the source, through the load, then the control device, and back to the other side of the source. This flow of electrons moves from the negative side of the source through the rest of the circuit toward the positive side of the source. Notice that a series circuit has two wires connecting the source to the load: one to send the electrons to the load, and the other, a return path, to bring them back to the source. During this process, some of the electrons are used up as they are converted to light, heat, sound, and so on.

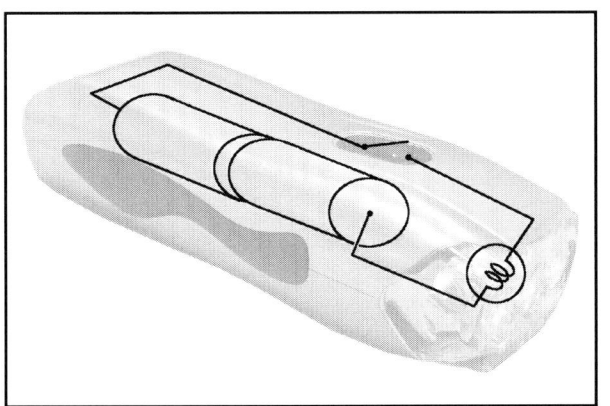

A flashlight is an example of a series circuit and is also a simple DC circuit. To represent such a circuit, a pictorial diagram may be used. A method that is preferred by electricians and technicians is to use a schematic diagram with interconnected symbols that each represent different electrical components.

The source of electrical energy in a flashlight is two series-connected dry cells, each having about 1.5 volts, which supply 3 volts to the circuit. A light bulb is the load, and a switch is connected between the source and load. The conducting path is supplied by a metallic strip along the inside of the tube holding the batteries. When the switch is open, no current flows and the lamp is off. When the switch is closed, a complete path exists and current flows through the circuit, lighting the lamp.

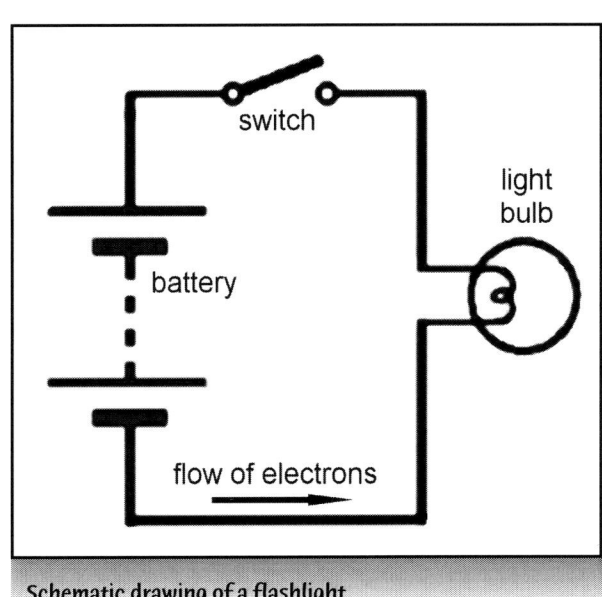

Schematic drawing of a flashlight

18 ■ Television Technical Theory

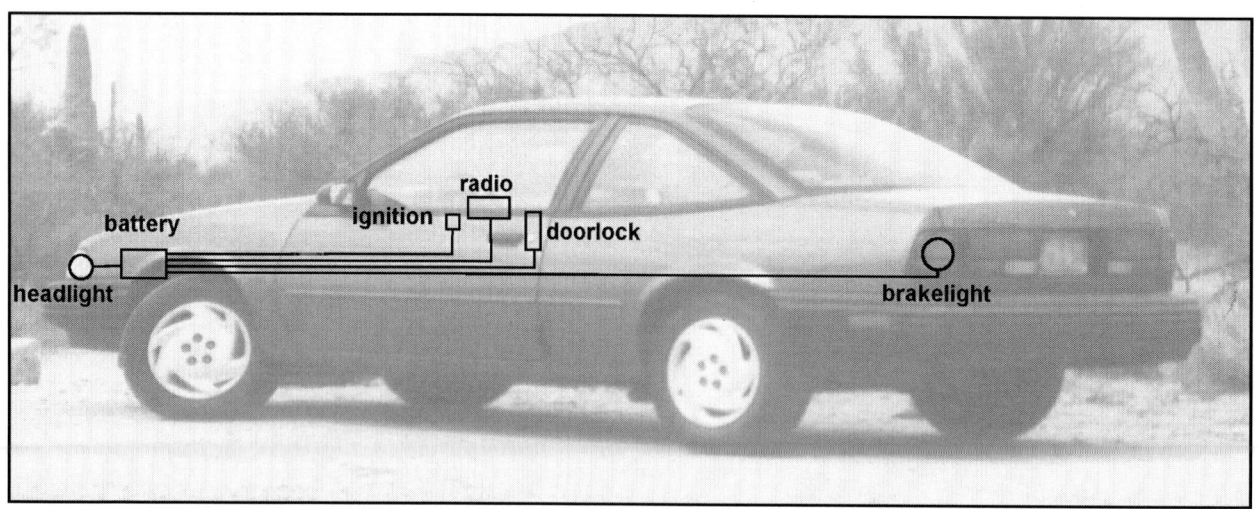

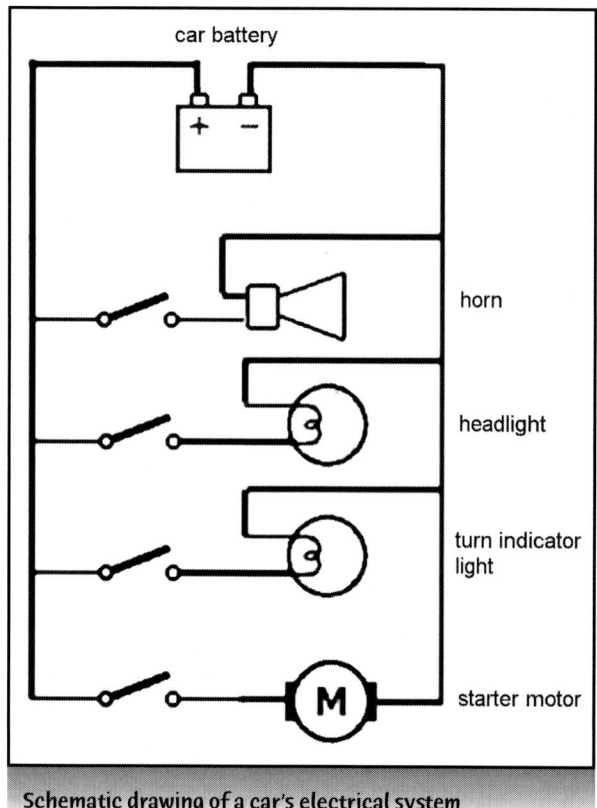

Schematic drawing of a car's electrical system

Parallel Circuits

A parallel circuit is one in which all the loads work at the same voltage as the source and independently of one another. If one load is switched off, the rest are unaffected.

The electrical system in an automobile is an example of a DC parallel circuit in which the 12 volts from the battery simultaneously supply electrical energy for the ignition system, headlights, taillights, and radio.

If another load is added in a parallel system, it supplies another path for the current; so the total current needed from the source increases. The combined resistance of the parallel circuit effectively decreases whenever another load (resistance) is added in parallel. As in a series circuit, the total power in a parallel circuit is the sum of the individual powers required.

Why Are You Being Told All of This?

Because . . . practically everything in a television station runs on electricity. Virtually everything on the technical side of television is a circuit. The concept of a circuit path is one that should be kept in mind when dealing with all the station's functions. Remember, all electronic television equipment not only needs power to operate, but also needs a signal path through which it sends and/or receives video, audio, or computer signals.

All AC wiring in a television facility is a network of series and parallel circuits. If you plug a light into the wall, you've created the simplest of series circuits. If you plug two lamps into the same duplex wall outlet, you've created a parallel circuit—made up of two series circuits, of course.

HOW IT ALL CONNECTS TOGETHER

"Sattinger's Law: It works better if you plug it in."
—Arthur Bloch, *Murphy's Law and Other Reasons Why Things Go Wrong*

Connectors, Patch Bays, Routing Switchers

In a typical television station, one is constantly surrounded by consoles with phosphorescent screens, small coloured indicator lights and meters, knobs, faders, levers, switches, and push buttons. Often, next to this equipment, there are racks of intimidating patch cables, routing gear, and mysterious black boxes. To the uninitiated, just looking at a patch bay can give one a sense of hopelessness, or at the very least, eyestrain.

Most production facilities and stations have several kilometres of wire within their walls and ceilings. These cables carry video, audio, time code, digital computer data, or control signals from one device to another. Some of the wires in a television facility go from the machines to a patch rack, and some of those then go to another patch rack in another area. Others go to a patch rack, and then to a routing switcher, which is nothing more than an oversize, semiautomatic patch panel. We'll explore all of these paths in this chapter.

In order to create television productions, we assemble the machines in some useable form using all of these interconnections. There are many correct ways to assemble the equipment—some are easier than others. There are also some incorrect ways to assemble the equipment. This chapter will explore some of the things to consider when hooking up all of this gear.

Most of what we use in a studio environment is already connected together for us in a standardized fashion. But for flexibility, and to make maximum use of all of our expensive equipment, practically all of those standard connections go through patch bays so that custom configurations can be arranged, and existing connections interrupted, at will.

First, let's begin with the basic connections used for audio and video in a broadcast production facility.

Audio Connections

As you read in the "Electricity 101" chapter, normal series circuits require two wires—one to supply the electrical energy to a device, and another that returns the unused current back to the source.

Most home (consumer) audio equipment sends the signal down a single wire and returns it via the common "ground" or "shield" line. The cable is usually terminated in a connection called an RCA plug, named after the Radio Corporation of America. The plug fits into a receptacle called, not surprisingly, an RCA jack. This system creates simple series circuits, and, for short cable runs, is quite economical and works

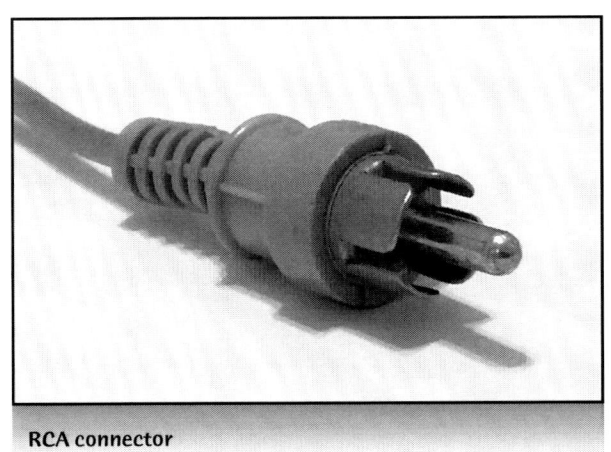
RCA connector

well. It's called an "unbalanced" connection. These are the sorts of hookups you make when you are connecting your VCR or DVD player to your television set at home, for example. This system is often used for both audio and video in these setups.

However, broadcast audio requires three wires for each audio connection. Why do we do this? Often, audio signals must be run considerable distances (across studio floors, all around a television station, etc.). And, usually running alongside these audio cables are sources of high voltage and current, for example AC wiring to power the equipment. When two cables run side by side, larger currents from the AC power cable, alternating at 60 cycles per second, induce small currents into the audio line. This is called electromagnetic induction.

Electrical levels in an audio cable generally are around a fraction of one volt. Therefore, if even a tiny little bit of AC current is introduced into an audio cable, we will be able to hear this additional signal at our receiving equipment (e.g., audio mixer, recorder) as a low-frequency humming sound at 60 Hz. Most of us are all too familiar with this annoying sound in audio recordings. We solve this problem in a rather unique way. To understand this thoroughly, we have to look at a particular characteristic of electrical sound signals.

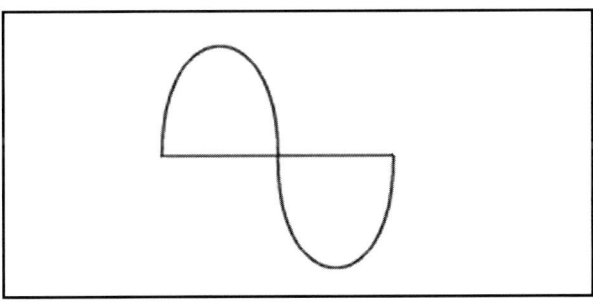

Phase

We illustrate electrical sound signals using pictorial representations of voltage fluctuations. Here we see a picture of one cycle of a sine wave sound. The positive "bump" represents a compression of air molecules in the original sound, turned into an electrical signal, and the negative part shows a rarefaction (expansion.)

This wave, as illustrated, has another interesting characteristic. It starts at "zero," moving slowly positive to a peak, then back through zero, then towards a negative peak, then back to zero. We've assumed in this case that the sound source starts making audio at a nominal "no expansions or compressions," or zero crossing, point. But there's no law saying that sound sources always have to do this.

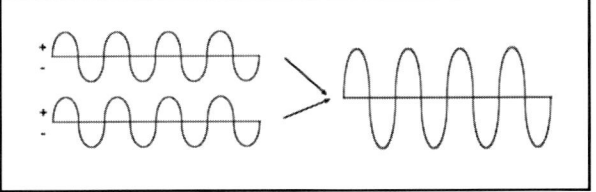

Let's now look at two electrical sound sources with exactly the same frequency, both starting at zero crossing points. If they are mixed together (by, say, an audio mixer) after they have travelled through their respective circuits, they will, in effect, "add up," making the final sound appear louder to us. This seems reasonable. After all, two identical sound sources, when mixed together, should be louder than just one.

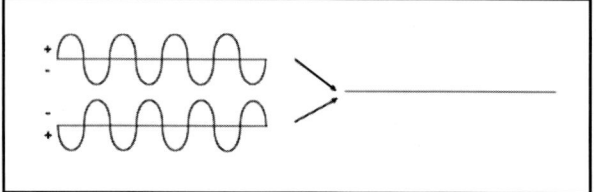

Now look at another pair of sound signals. This time, one of them begins positive, but the other one goes negative. When these sources are mixed together, the positive voltage swings from one source will combine with the negative voltage swings of the other source, essentially cancelling each other out. What we will hear is not a louder sound. In fact, we will hear nothing at all! This is an example of "phase cancellation." We say that the first signal is "180 degrees out of phase" with the second one.

Given this interesting electrical principle, suppose we try sending two "versions" of our audio from one place to another. One of these is the "normal" one, involving a signal wire and a return wire (also called a "ground"). That takes two wires, and is the same setup as a regular series circuit. In broadcast audio, we also send another version of the signal, using an additional wire to send it, and using the same ground wire as the first signal for its return path. That's a total of three wires within the audio cable. The second version of our audio signal, however, is 180 degrees out of phase with the first one. This means that it's electrically "upside down" from the first signal—when a voltage goes positive on the first signal, the voltage on the out-of-phase line goes negative by exactly the same amount and vice versa. Think of the two signals as electrical mirror images of one another.

What's the advantage of this? Broadcast equipment is designed to take both versions, and, deep inside the recording equipment or mixer, for example, invert the phase of the second signal so that it's back in phase with the first and then mixing

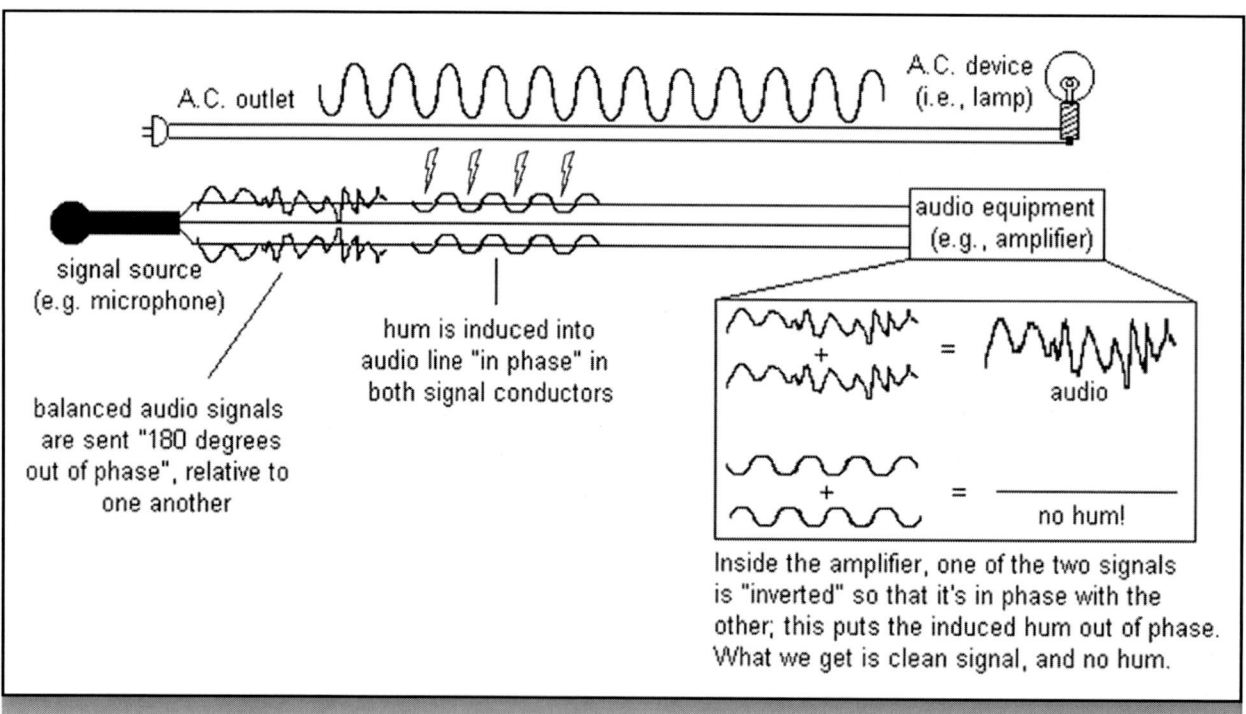

Why we used balanced audio in the broadcast world

both signals together. What will happen then? For the audio signal part of the electrical current, not much—you get a little more level, a little louder sound, than before. But, if a hum has been induced into both signal lines equally (and in phase), it will travel to the receiving equipment. One of the hums will be inverted out of phase and mixed with the first one inside the equipment itself, and the hum part of the signal will be cancelled out!

Therefore, in broadcast audio circuits, the audio signal travels down two of the three wires, with the last one being a common return signal wire for both of these signals. This is called a "balanced" line, and differs from most home audio equipment, which sends the signal down a single wire with one return wire. If you look at common broadcast connectors, you will see the three connections that have to be made, as opposed to the two used in home equipment. Audio patch cords also have all three wires in them.

XLR connector

Professional balanced connections are used in all analog audio cables including microphone lines, CD players, digital audio recording devices, and so on. These professional terminations are called XLR connectors. Originally produced by Cannon as the "X" series of connectors, they later added a latch ("L") so the connectors wouldn't pull apart accidentally. Finally, a surface of rubber ("R") surrounds the pins or holes in these connectors.

Consumer Equipment in Professional Applications

You can use consumer equipment in professional applications. But you have to convert from one connection system to the other. Normally this is done by using a balancing amplifier which takes the consumer equipment's unbalanced output, adjusts for the level differences between consumer and professional gear, and creates the two out-of-phase signals. XLR connections, on the output of this amplifier, interface to regular broadcast equipment.

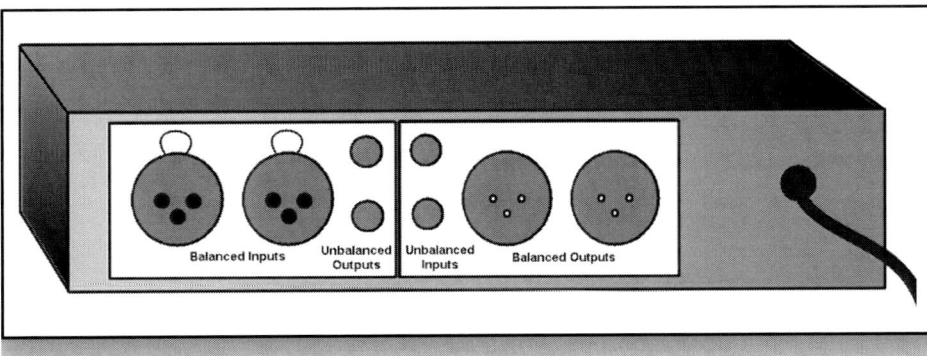
Balancing amplifier

Video Connections

Broadcast video requires two connections for each video line, just the same as consumer video equipment. The signal travels down a single wire with the "shield" or ground as a return path. If you look at common broadcast video connectors, you will see the two connections that have to be made.

BNC connector

The connector used for video is called a BNC (Bayonet Neill-Concelman.) This is named after its bayonet locking system and its two inventors, Paul Neill of Bell Labs and Amphenol engineer Carl Concelman (inventor of the C connector). You may also see it referred to as a Baby Neill-Concelman, Baby N connector, British Naval Connector, Bayonet Nut Connector, or Bayonet Naval Connector, although these terms are, strictly speaking, incorrect.

Making the Connections

It would be an awkward situation if every time you wanted to use a piece of equipment, you had to crawl around behind the machine and plug in wires to connect it to the system. So, most equipment is connected together in normal usage configurations by engineering systems designers. However, if we wish to hook up a piece of equipment in a way not originally envisioned by the designers, we need a way of breaking the original connections to make new ones of our own. Patch racks allow us to do this. In any typical technical environment you will probably find both audio and video patch racks. These connect the nearby equipment in one central area to allow us to make different connections. Frequently, the patch racks are configured in such a way that if there are no patch cords in the system, equipment will ordinarily be connected ("normalled") to other equipment to accommodate the way in which we will most frequently be using it. Usually, you will see two rows of jackfields. A common convention followed is "out to in," meaning that the equipment's outputs are on the top row of a double jackfield, and the equipment's inputs are on the bottom row. This makes normalling a straightforward and logical process inside the patch rack. To reconfigure the existing wiring, you insert a patch cord from one machine's output, to a second piece of equipment's input. It's that simple. The same procedure is used for both audio and video patch racks.

Patch rack

TRUNK LINES

If all the equipment were in one central area, there would be no need for trunk lines. The equipment's connections would come to one large patch bay and would be interconnected there. But most facilities have equipment in several different areas in the building. Trunk lines are the next logical step. They are nothing more than a series of audio, video, or computer data cables connecting one area to another. They're just wires running through a production facility's walls and ceilings.

Distribution Systems

Frequently, we're required to send a single feed to two or more different places. At first glance, it seems to be a simple matter of gathering together the source machine's output cable, along with all of the destinations, and connecting all of the circuits together into one big connection. But electronics and physics being what they are, you don't get something for nothing; because of the way broadcast systems are constructed, every time you add another piece of equipment's input to this primitive distribution system, you reduce the signal available to all the destinations. There is a solution to this, and it is used throughout the station.

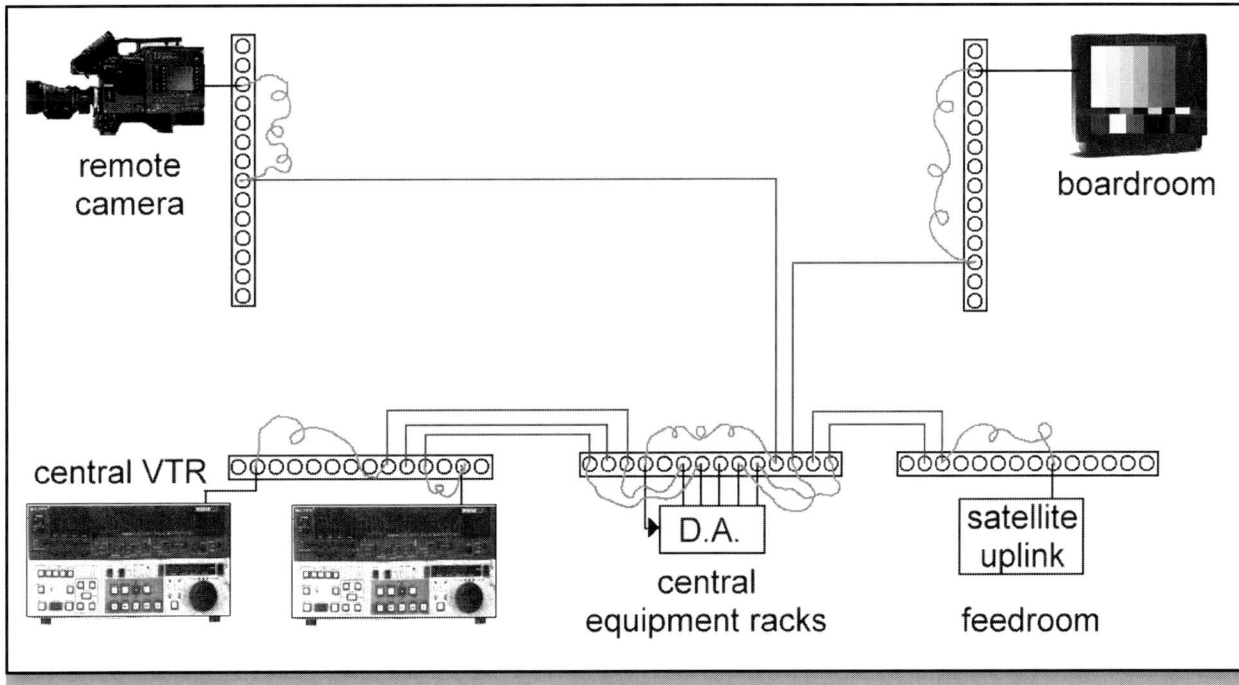

Example of the use of video trunk lines in a television facility

DISTRIBUTION AMPLIFIERS

These units take a single input and pass it through four or six individual amplifiers, which provide us with separate, isolated outputs. Those outputs don't load each other down. Whenever a signal is distributed to more than one place, a distribution amplifier (or DA, as it's commonly called) is used. There are different kinds of DA's available for audio and video. Some audio distribution amplifiers are stereo, which simply means that there are two single channel units in one package. So, if you're sending a source to more than one place, you will need to use the DA's available at your patch racks.

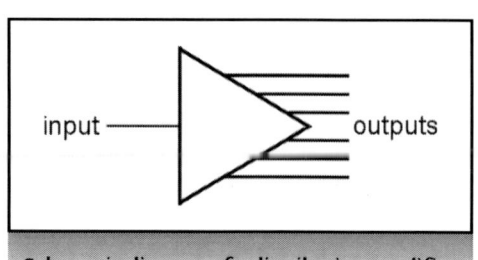

Schematic diagram of a distribution amplifier

Audio Phasing

We read earlier about how audio signals can be in phase or out of phase relative to one another. A similar problem can occasionally occur in audio distribution systems.

Normally, multichannel audio signals are mixed together and laid down on audio tracks in phase. If a sound appears on two channels, and those channels are later mixed together, the sounds should "add up." If, for some reason, one channel is electrically out of phase with respect to the other, the sounds cancel, as we have already seen.

How could one channel ever get out of phase with another in the first place? Consider that broadcast audio is a balanced system: two wires carry signals, and a third is an electrical return connection. If somewhere in all the audio interconnections, the two signal wires are reversed in any one channel (perhaps by accident during the wiring of a production facility), that signal will now be out of phase relative to all other channels. The obvious way to correct an electrical out-of-phase condition is to rewire the equipment properly. If there is no time for that, there are two other temporary solutions.

Most audio consoles have a phase reversal switch that shifts a channel's phase by 180 degrees, and therefore temporarily back into phase. However, if you are not using an audio console at the time (for example, during a direct transfer from one recording device to another), you can also use a special patch cord, generally referred to as a "phase reversal cable," which has those two wires switched over within its connections. Therefore, with this cable, an out-of-phase channel can be put back in phase with

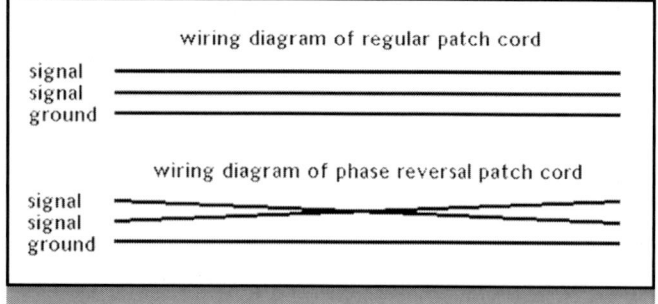

Schematic diagram of 'normal' and 'phase reversal' patch cords

respect to its accompanying channel. However, if the two channels are already in phase and the phase reversal cable is used accidentally, one of the channels will then be put out of phase, causing problems! As you can see, this is a very useful cable, but a potentially dangerous one as well.

Routing Switchers

As mentioned before, routing switchers are large, automated patch bays. As the output from each device enters the routing switcher, it is put through many distribution amplifiers—one for each place to where it can possibly be sent by the router. The internal switches in the router are controlled by circuits, which, in turn, are connected to routing switcher control panels throughout the building. When you select a destination, and a video or audio source that you'd like put there, the signal is switched within the routing matrix automatically. Routing switchers are designed for both audio and video connections.

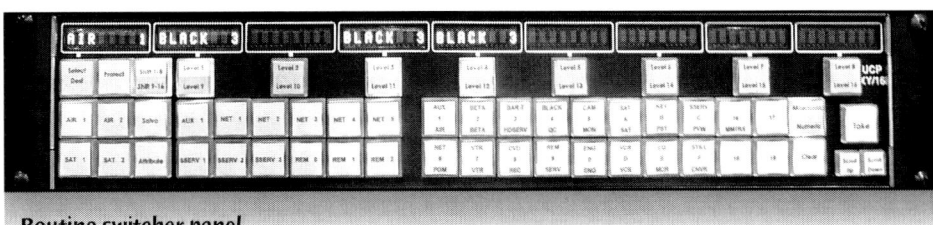

Routing switcher panel

Intercom Systems

Connecting the people involved in a teleproduction is just as important as connecting all of the equipment together. The generally accepted way to speak to other parties involved in a television operation (apart from the frantic screaming sometimes witnessed during an intense moment in the control room) is through an intercom panel.

Intercom systems come in various sizes, shapes, and complexities, but the basic principle is the same in all of them: transmit someone's voice to some other particular person, with no distortion or interruption from other parties. Sometimes many parties may need to speak to one individual. At other times, one person wants to speak with a whole group of people at once.

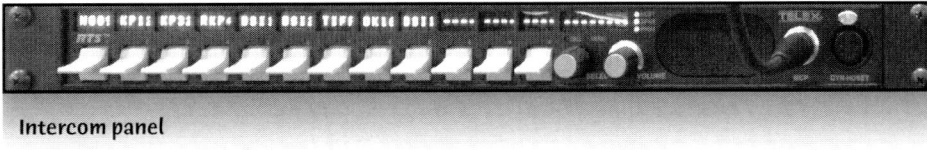

Intercom panel

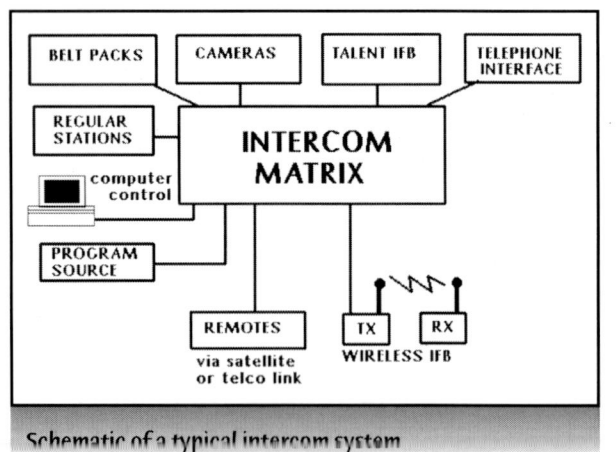

Schematic of a typical intercom system

The standard intercom system will involve several intercom base stations (the ones mounted in control room consoles and equipment areas); floor director belt packs; camera headsets; on-air talent IFB (an Interruptible Fold Back earpiece); a source of on-air program audio that's often fed to the on-air talent; and some type of computerized control of the whole system. As well, more complex systems provide the ability to interface with someone on a telephone outside of the production facility; satellite intercom channel communications from a remote feed; wireless talent IFB; wireless floor director systems; and interconnection with two-way handheld radios.

Data Circuits

Beyond the usual video and audio connections, there are many other control signals being sent around the station. Some of these interface with character generators and switchers; others start and stop video recorders by remote control; still others steer satellite dishes.

The easiest way to think about all of this data transmission is to break it down into three parts: a transmitter of information, the information path itself, and a receiver of the data. Often, the same piece of gear will transmit and receive signals. These systems don't use human language; they have their own specific computer codes to tell each piece of equipment what to do.

Terminals

Terminals are the human interface to the machinery. These can be in almost any configuration. For example, the transmitter controls and alarm panels in a master control transmission facility are a kind of terminal—this is occasionally a panel with indicator lights, although in other installations it's a computer with special software to monitor the transmitter status. The routing switcher controllers and production switcher panels are terminals, as are the intercom stations. All of these terminals send commands to other pieces of equipment. Some terminals can even be called up on the telephone and will speak to you; by punching in a series of numbers using a touch-tone phone, you can communicate with it, ask it questions about the status of equipment, get verbal responses, and even effect changes in the system.

Talking to Machines

Computers and other terminal devices communicate with each other using only electrical signals—changes in the current or voltage on a wire. Because computers can only sense the absence or presence of a voltage within an electrical system, the binary number system is used. The "0" and the "1" are the symbols of the binary system; with computers the "0" represents an absence of a voltage and the "1" represents a presence of a voltage. One of these binary digits is commonly referred to as a bit.

In the numbering system mostly commonly used by people (base ten), we have the digits zero through nine to work with. In a number such as 345.27, we understand it to mean "three hundreds, plus four tens, plus five ones, plus two tenths, plus seven hundredths." Similarly, in the binary system, the columnar position of the numbers denotes their "weight" within the final number. Each additional column within the binary system is worth two times what the previous column was.

Counting up to eleven in the decimal system is done like this:

```
 1 - one    "ones"
 2 - two    "ones"
 3 - three  "ones"
 4 - four   "ones"
 5 - five   "ones"
 6 - six    "ones"
 7 - seven  "ones"
 8 - eight  "ones"
 9 - nine   "ones"
10 - one    "tens", no  "ones"
11 - one    "tens", one "ones"
```

Counting up to eleven in the binary system, where we only have two symbols, looks like this:

```
   1 - one "ones"                                              = 1
  10 - one "twos", no  "ones"                                  = 2
  11 - one "twos", one "ones"                       = 2+1      = 3
 100 - one "fours", no  "twos", no  "ones"                     = 4
 101 - one "fours", no  "twos", one "ones"         = 4+1      = 5
 110 - one "fours", one "twos", no  "ones"         = 4+2      = 6
 111 - one "fours", one "twos", one "ones"        = 4+2+1    = 7
1000 - one "eights", no  "fours", no  "twos", no  "ones"       = 8
1001 - one "eights", no  "fours", no  "twos", one "ones" = 8+1  = 9
1010 - one "eights", no  "fours", one "twos", no  "ones" = 8+2  = 10
1011 - one "eights", no  "fours", one "twos", one "ones" = 8+2+1 = 11
```

What can be done with these binary numbers? We can certainly do arithmetic with them, and all we need is an absence or presence of a voltage.

It's also a very straightforward process to convert the letters of the alphabet and other special symbols into codes that can be represented by numbers. If we take the binary numbering system, and use eight columns (bits), we have a possible 256 combinations (2^8) to choose from. This is more than enough to represent letters, numbers, and special symbols. In fact, a common system called ASCII uses 8 bits to represent these elements, and is commonly used in computers in most of the world. These are the codes sent to and from monitoring terminals connected to a piece of equipment.

30 ■ Television Technical Theory

000		(nul)	016	^P	(dle)	032	sp	048	0	064	@	080	P	096	`	112	p
001	^A	(soh)	017	^Q	(dc1)	033	!	049	1	065	A	081	Q	097	a	113	q
002	^B	(stx)	018	^R	(dc2)	034	"	050	2	066	B	082	R	098	b	114	r
003	^C	(etx)	019	^S	(dc3)	035	#	051	3	067	C	083	S	099	c	115	s
004	^D	(eot)	020	^T	(dc4)	036	$	052	4	068	D	084	T	100	d	116	t
005	^E	(enq)	021	^U	(nak)	037	%	053	5	069	E	085	U	101	e	117	u
006	^F	(ack)	022	^V	(syn)	038	&	054	6	070	F	086	V	102	f	118	v
007	^G	(bel)	023	^W	(etb)	039	'	055	7	071	G	087	W	103	g	119	w
008	^H	(bs)	024	^X	(can)	040	(	056	8	072	H	088	X	104	h	120	x
009	^I	(tab)	025	^Y	(em)	041	)	057	9	073	I	089	Y	105	i	121	y
010	^J	(lf)	026	^Z	(eof)	042	*	058	:	074	J	090	Z	106	j	122	z
011	^K	(vt)	027	^[	(esc)	043	+	059	;	075	K	091	[	107	k	123	{
012	^L	(np)	028	^\	(fs)	044	,	060	<	076	L	092	\	108	l	124	|
013	^M	(cr)	029	^]	(gs)	045	-	061	=	077	M	093	]	109	m	125	}
014	^N	(so)	030	▲	(rs)	046	.	062	>	078	N	094	^	110	n	126	~
015	^O	(si)	031	▼	(us)	047	/	063	?	079	O	095	_	111	o	127	⌂

128 Ç	144 É	160 á	176 ░	192 L	208 ╨	224 α	240 ≡
129 ü	145 æ	161 í	177 ▒	193 ⊥	209 ╤	225 ß	241 ±
130 é	146 Æ	162 ó	178 ▓	194 ┬	210 ╥	226 Γ	242 ≥
131 â	147 ô	163 ú	179 │	195 ├	211 ╙	227 π	243 ≤
132 ä	148 ö	164 ñ	180 ┤	196 ─	212 ╘	228 Σ	244 ⌠
133 à	149 ò	165 Ñ	181 ╡	197 ┼	213 ╒	229 σ	245 ⌡
134 å	150 û	166 ª	182 ╢	198 ╞	214 ╓	230 μ	246 ÷
135 ç	151 ù	167 º	183 ╖	199 ╟	215 ╫	231 τ	247 ≈
136 ê	152 ÿ	168 ¿	184 ╕	200 ╚	216 ╪	232 Φ	248 °
137 ë	153 Ö	169 ⌐	185 ╣	201 ╔	217 ┘	233 Θ	249 ·
138 è	154 Ü	170 ¬	186 ║	202 ╩	218 ┌	234 Ω	250 ·
139 ï	155 ¢	171 ½	187 ╗	203 ╦	219 █	235 δ	251 √
140 î	156 £	172 ¼	188 ╝	204 ╠	220 ▄	236 ∞	252 ⁿ
141 ì	157 ¥	173 ¡	189 ╜	205 ═	221 ▌	237 φ	253 ²
142 Ä	158 ₧	174 «	190 ╛	206 ╬	222 ▐	238 ε	254 ■
143 Å	159 ƒ	175 »	191 ┐	207 ╧	223 ▀	239 ∩	255

ASCII computer codes

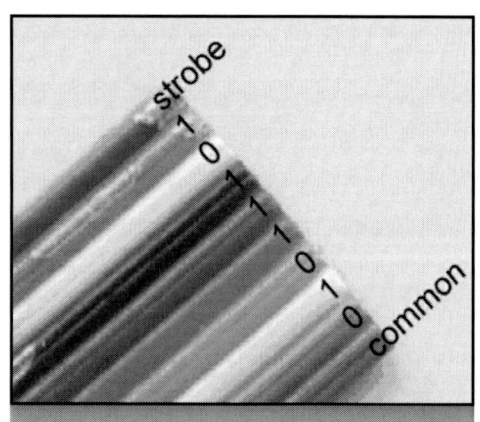

Parallel transmission: needs ten wires

Transmission of Computer Signals

Now that we know how to count and talk like a computer, we need some way of getting our communications across to the machines. There are two ways of doing this.

Parallel

In parallel transmission, each bit of a character travels on its own wire. An additional wire, carrying what is called a strobe, indicates to the receiver when all of the bits are present on their respective wires so that the binary values can be read. So, to send eight bits plus a strobe, and some form of common return for all of the other signals (remember, these are all individual circuits) requires at least ten wires. This system is fine for short distances, but because of the cost of all that wire, it becomes prohibitively expensive to send a signal over many metres or kilometres.

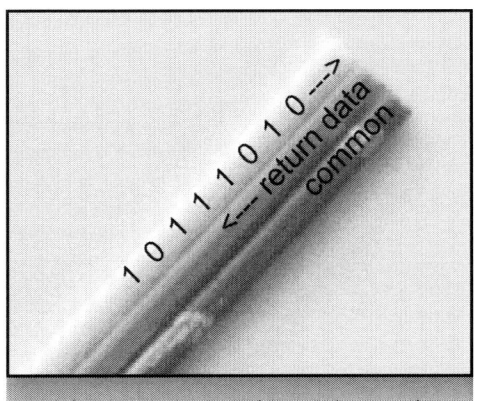

Serial transmission: needs two wires, or three for full two-way transmission

Serial

Serial transmission is used for longer distance communications. Each bit is sent, one by one, down a single wire. Eight of these bits (one full character) is called a byte. Before the beginning of each eight-bit byte, a "start" bit is sent; after the character's eight bits are sent, a "stop" bit is then transmitted. This tells the receiving equipment where the beginning and end of each byte is located, relative to the continuous stream of information.

Modem

Modems

Modems are simply a way of sending the serial chain of on/off sequences down a telephone line that normally expects audio frequencies. The modem is a unit that converts the bits into a series of audible tones that differ depending on whether they're representing the "1"s, or the "0"s. This is a called "modulating" the data. At the receiving equipment, these tones are converted back into the original electrical signals. This is known as "demodulation." The word "modem" is an abbreviation of "modulator/demodulator." More expensive and complicated modems allow transmission of computer data at very high speeds, using various frequency- and phase-shifting methods.

Fibre Optics

Instead of using wires, it's possible to use the on/off electrical sequences to turn a laser beam on and off. The resulting modulated light beam can be sent down a very small tube of glass fibre along great distances. At the other end, the flashing light beam can once again be converted into electrical on/off sequences. This is the principle behind fibre optics. The light doesn't come out of the sides of the fibre, due to the principle of refraction. Within a fibre optic cable, there is a centre core of glass fibre surrounded by a coating (or cladding). The core has a refraction

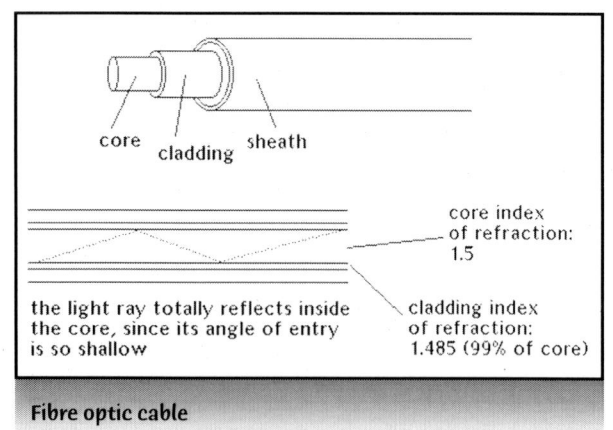

Fibre optic cable

index that is slightly more than the cladding. When a light ray enters the glass fibre, it may not be coming in exactly parallel to the edges. It will approach the cladding next to the fibre, but, because the cladding index is slightly less than the glass, it gets reflected back into the glass again, and so on, down the length of the fibre.

You can see this effect by trying a simple experiment. Fill a glass half full of water, and place a spoon in it. When you look through a glass of water at an object, it doesn't appear to be exactly where it ought to be. This is because the light rays have been bent as they go through the water; the speed of the light rays increases as they leave the water. The water has a refraction index that is different from the air. This "total internal reflection" property can also be illustrated with your glass of water: simply look inside the top of the glass, and you will see your own reflection on the far side of the tumbler.

INTERFACE STANDARDS

All of the pieces of equipment around a television facility that use data connections have different types of connectors and different ways of talking to one another. Terms like RS-232C and RS-422 abound.

RS-232C is a serial interface standard and is commonly used for connecting computers to modems and serial printers, usually employing a 9 or 25 pin connector. It uses large voltage swings (typically ±12 volts, but may be as high as ±25 volts) to send the serial data. All of the signals are referenced to a single common return (ground). This is fine most of the time, but sometimes the RS-232C system runs into problems when the equipment at either end of the transmission is plugged into a different AC outlet or circuit.

RS-422 attempts to overcome that problem by using two wires for each signal and sensing the voltage difference between these lines to represent the two logic levels. Because of this, the interface can also use much lower voltage swings to represent the data—typically within ±5 volts.

5

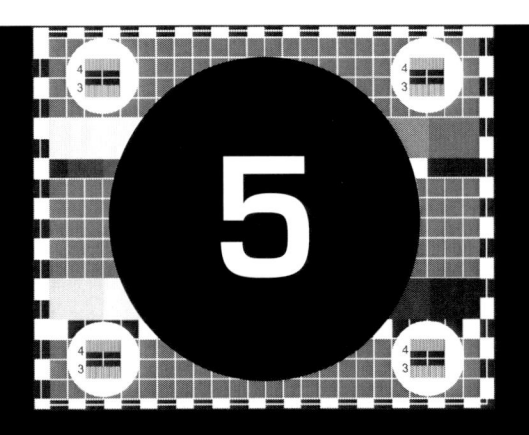

ANALOG VIDEO

"If I can't picture it, I can't understand it."
—Albert Einstein

Vision

Let's think about human vision for a moment. Our eyes receive an image, and hundreds of thousands of fibres in the optic nerve simultaneously send to the brain small electrical signals that, taken together, represent the whole scene. Human vision, therefore, uses an abundance of "channels," all at once.

In television, however, the entire scene must be sent through a single channel. Think of it as a serial process, sent down a series circuit. Within the camera an electrical signal is formed to represent the changing brightness and colour of each area of the scene. This signal is sent to the monitor. At the monitor the signal is transformed back into light, and the image is assembled on the viewing screen in its proper relative position.

In the television system, the picture we want to see is "scanned" sequentially, top to bottom, left to right. A full frame of video is scanned in approximately 1/30th of a second, so we say that television runs at a rate of 30 frames per second.

Even though the picture elements are laid down on the screen one after the other, they all must be perceived at once. This requirement is met by persistence of vision, a property of our eyes and brains. When light entering the eye is shut off, the impression of light persists in our brain for about a tenth of a second. You can see this yourself; if you close your eyes, then open and shut them quickly, you will still "see" an image in your brain for a brief moment after your eyes are closed.

So, if all the picture elements in the image are presented successively to the eye in a tenth of a second or less, the whole area of the screen appears illuminated to us, although in fact only one spot of light is present at any moment in time. Activity in the scene is represented, as in motion pictures, by a series of still pictures, each differing slightly from those preceding and following it.

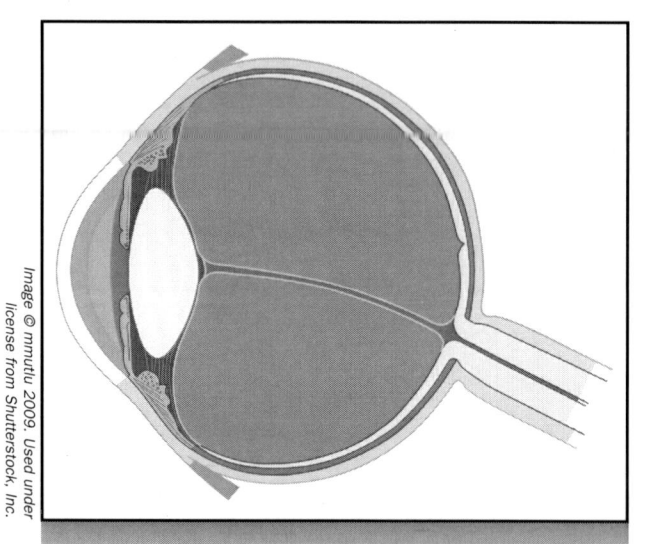

Cross-section of human eye, showing optic nerve

The perception of motion comes to us by a series of still images

BLACK-AND-WHITE TELEVISION

Vladimir Zworykin and Philo Farnsworth both developed electronic television using the same principle: that a narrow stream of electrons could be manipulated to both capture and reproduce moving images. In this section, we will investigate how electron beams are controlled to create a black and white television image.

Electrons

Before we go further in our discussion of video, let's take a moment to have a look at what electrons are.

The electron is often described as a "particle of electricity." Electrons have a negative electric charge. They were discovered in 1895 by Joseph J. Thomson, a British physicist, in the form of cathode rays—streams of electron particles. What's interesting about cathode rays is that they can be attracted and repelled by both magnetic and electric fields. Because electrons are essentially weightless—they have a mass of 9.1083×10^{-28} grams—this is fairly easy to do. If we move the stream of electrons around in a particular pattern, we get what is known as scanning.

The principle of scanning (note how each line breaks down the scene into discrete elements of picture intensity)

Scanning

The process of breaking down the scene into picture elements and reassembling them on the screen is known as scanning. The pattern is very similar to what your eyes do when they read a book—they move left to right across a line of text, as you slowly work your way toward the bottom of the page. In television scanning, the scene is broken up into a series of horizontal lines of information.

At the Camera

The camera "reads" the topmost line of scenic information from left to right, producing a series of electrical signals that corresponds to the light and shadows along that line. Once it gets to the right side of the picture, it returns back quickly

36 Television Technical Theory

One line from the Parthenon shot

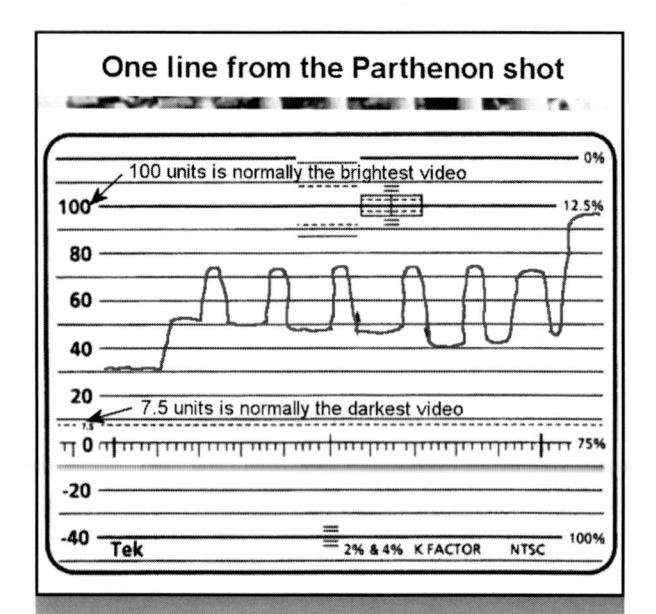

A voltage created as the camera scans across one line of the Parthenon shot

to the left, on the next line below, and follows it in the same way. In this way the camera reads the whole area of the scene, line by line, until the bottom of the picture is reached. Then the camera scans the next image, repeating the process continuously. The television camera has now produced a rapid sequence of electrical impulses; they correspond to the order of picture elements scanned in every line of every image.

At the Monitor

At the television monitor this signal is recovered and controls the picture display. The monitor creates an image that is composed of horizontal lines just like those produced in the camera. As the camera examines the topmost line, a spot of light is simultaneously produced in the monitor, and as the spot moves from left to right, it reproduces the topmost line of scene information on the screen. The video signal causes the spot of light to become brighter or darker as it moves, and so the picture elements scanned by the camera are reproduced line by line at the monitor, until the whole area of the screen is covered, completing the image. The process is repeated over and over again, at a high speed.

Normally the scanning lines are so close together that we don't see the lines—to our eyes, they merge together as one image. But if you look at a television monitor very closely (or with a magnifying glass) you can see the individual lines that make up the picture.

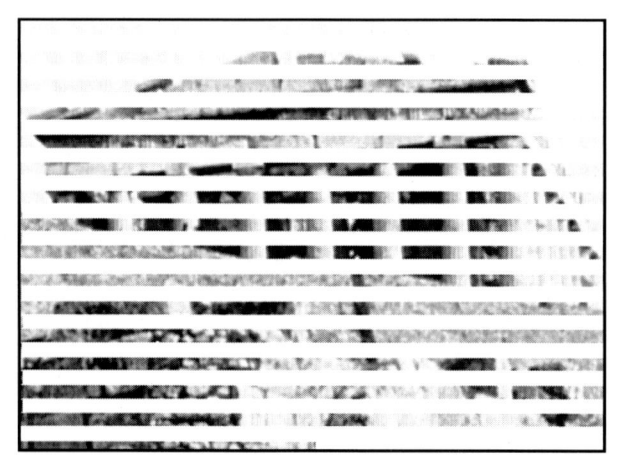

Interlace

In analog television using old-fashioned picture tubes, the phosphor coating on the face of the tube could only retain the picture information for a certain amount of time. As a result, the image on these early display devices would flicker if scanning from top to bottom only occurred at 30 frames per second. To avoid flicker, each still picture is presented using a process known as interlaced scanning. After the topmost line is scanned, space for another line is left immediately below it, and the next scanned line appears just below the empty space. As the scanning proceeds, alternate lines are scanned, with empty spaces between them. Once the scanning is at the bottom of the screen, only half of the full television frame information has been reproduced. The next top-to-bottom scan also consists of spaced lines, but its lines fall precisely in the empty spaces of the preceding image. When this has been completed, the screen has been filled by the two sets of interlaced scanning lines. Each of the scanning sets is called a field. There are two fields to each frame of television scanning, and each field therefore takes approximately 1/60 of a second to reproduce.

Synchronization Signals

The constantly changing voltage coming out of the camera is simply a varying electrical signal in a primitive form that represents the light and dark areas of a scene. These signals go from a low voltage (representing black) to a high voltage (representing white). They are measured in "video units." The video level of the darkest feature in a picture has been standardized at 7.5 units, and the brightest information in a scene should not normally be higher than 100 units. Video levels below 7.5 units are reserved for other purposes, as we will see in a moment.

The camera has to tell the monitor where there is a dark or light portion of a picture. But it also has to tell the monitor when to start and stop producing each line of video, and also when to begin and end each field as well. That's the purpose of synchronizing signals.

Scanning Left To Right, and Back Again: The Horizontal Interval

When a monitor is reproducing a line of video, it needs to be told when to begin that line, and when to finish, in synchronization with the camera. A group of special signals are sent by the camera to pass along this information. Once a line of picture information has been sent by the camera, the video falls below 7.5 units, to 0 units—well below the level of the darkest portion of the picture content. This tells the monitor to stop displaying video information to the viewer, and is called the "blanking" signal; it blanks out the picture content.

To ensure that a picture monitor's scanning doesn't drift off frequency over a long time (which would skew our picture in unpredictable ways), we send, in the middle of the horizontal blanking period, a "horizontal sync pulse." This gives the monitor a reminder to resynchronize with the camera's scanning at the end of every line. This sync pulse is at a level where it can easily be detected by the monitor, and will never be seen by the viewer: at −40 units on our video scale. The pulses will be sent 15,734 times a second (once for each line of video).

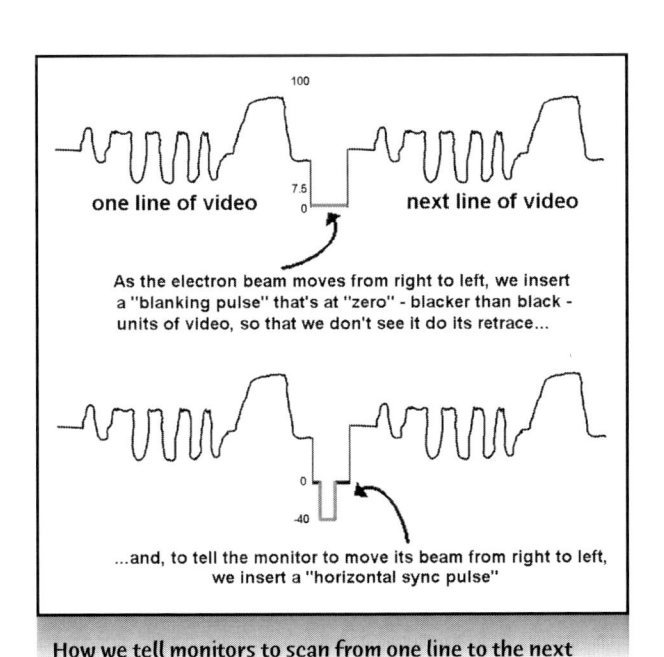

How we tell monitors to scan from one line to the next

As the electron beam moves from right to left, we insert a "blanking pulse" that's at "zero" - blacker than black - units of video, so that we don't see it do its retrace...

...and, to tell the monitor to move its beam from right to left, we insert a "horizontal sync pulse"

If there is a problem with the proper transmission of the horizontal sync pulses, or if the monitor isn't detecting and interpreting them properly, you get a result like the one in the picture (the original video picture is also shown here, to give you an idea of the effect).

No horizontal hold, due to missing horizontal sync pulses

Scanning Top to Bottom and Back Again: The Vertical Interval

A similar process occurs with the vertical sweep from top to bottom. During the vertical blanking period, a vertical sync pulse is created. This pulse tells the monitor "you'd better make your way back to the top of the screen now." The shape of the vertical sync pulse is actually six small pulses. It's made up this way to provide continuous synchronization for the horizontal scanning system of the monitor, even during the vertical retrace period.

In addition to the vertical sync pulse, another group of pulses is required when using interlaced scanning. Interlacing occurs because the second field of scanning starts half a line's distance across the screen, relative to the first field. The analog television system has 525 total lines per frame. Therefore, each field has half that many, or 262½ lines. Since 525 doesn't divide evenly by two (we end up with that extra half line), this means that one field must begin one half line later than the other one.

To make this rather complex process happen successfully, we put into the vertical interval signal one group of six "equalizing" pulses just before the vertical sync pulse, and a second group of equalizing pulses just after the vertical sync pulse. Basically, the equalizing pulses make sure that interlacing happens properly—that the scanning starts at the proper points in each of the two video fields.

Vertical roll due to lack of lock with vertical sync pulse

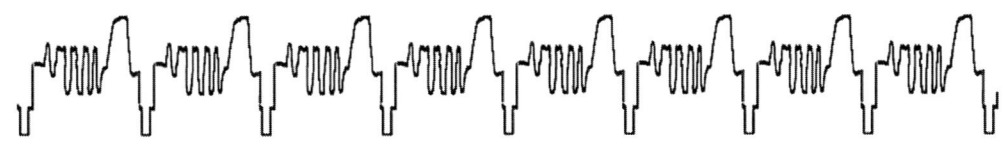

We keep generating line after line after line...241½ of them. Eventually, we are at the bottom of the screen, and have to get back up to the top.

To do this, we add some more "blacker than black" blanking, but this time it's for a much longer duration - 21 video lines are used for this vertical interval. After that, we start the second field of video, like this:

end of first field   start of second field

0 units

We have to tell the monitor to go back to the top. To do that, we insert a long sync pulse (3 video lines' duration), all the while continuing to tell the horizontal sawtooth to keep working.

This vertical sync pulse looks like this:

-40 units

Since we're making "interlaced" video, we have to tell the monitor to complete the first field with half a line of video, and begin with a full line of video for the second upcoming field.

We do this with "equalizing pulses" which sort of look like the vertical sync pulse, but upside down:

Finally, we've have a bunch of video lines left over, which are simply blanked (no peculiar pulses, but no video either.) These take up the remainder of the "vertical interval."

We then start the next field of visible video.

The vertical interval, featuring blanking, vertical sync pulse, and equalizing pulses

40 ■ Television Technical Theory

If you come across an older television set, you may find that it has two knobs called "horizontal hold" and "vertical hold." If yours has these, experiment with them. With a bit of fiddling with the vertical hold control, you can make the picture "roll" vertically down the screen. The "black bar" you now see in the middle of the screen is the vertical interval. If you play around with the horizontal hold control you may be able to cause the television set to lose horizontal lock as well. Modern televisions don't have these controls anymore, because synchronization technology in home receivers is now so reliable that we don't have to make these adjustments. If you have only a newer television and can't experiment for yourself, the illustrations here show you what vertical and horizontal intervals look like.

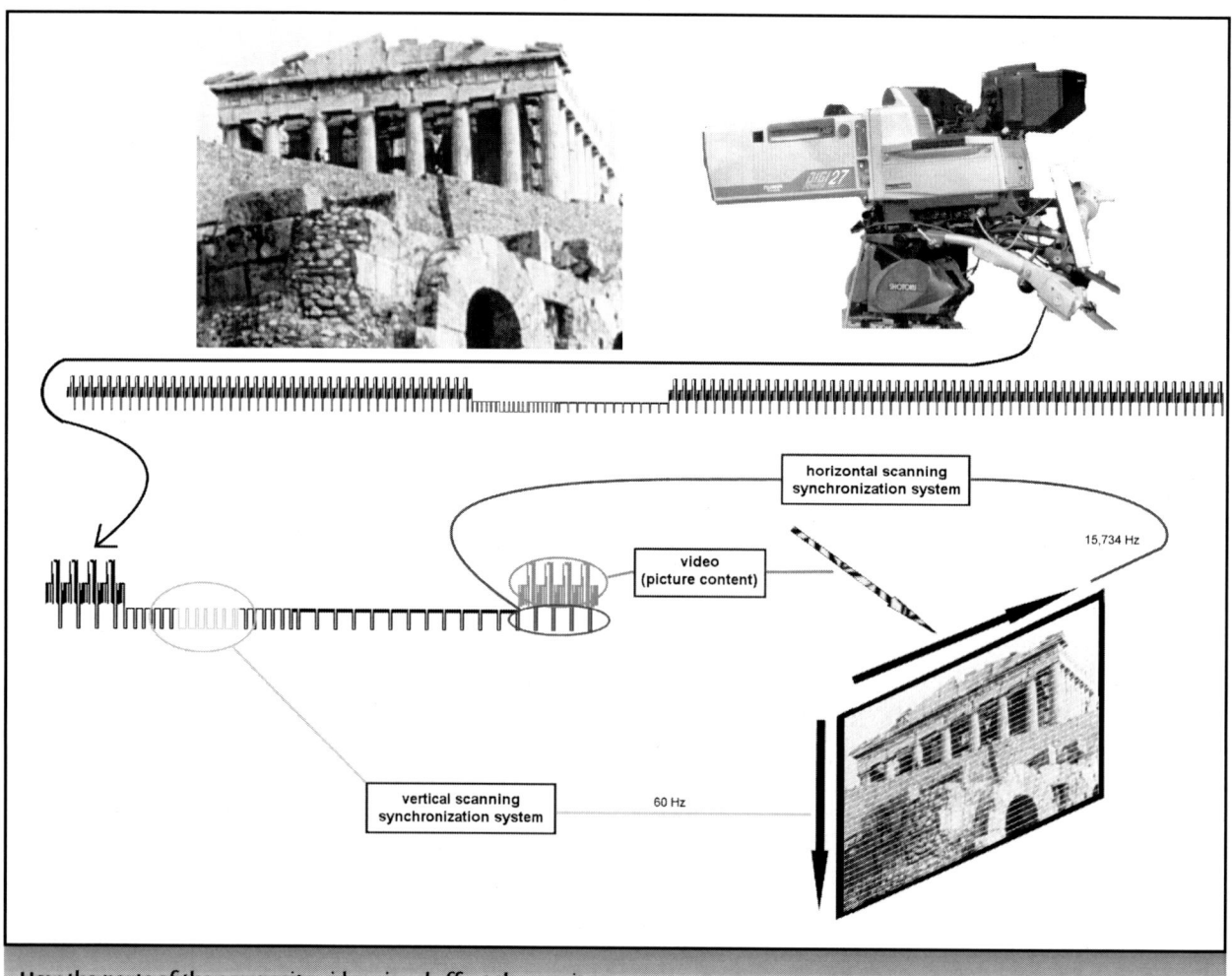

How the parts of the composite video signal affect the monitor

COLOUR TELEVISION

Colour television employs the basic principles of black-and-white television scanning. The essential difference is that a colour picture is like three pictures in one.

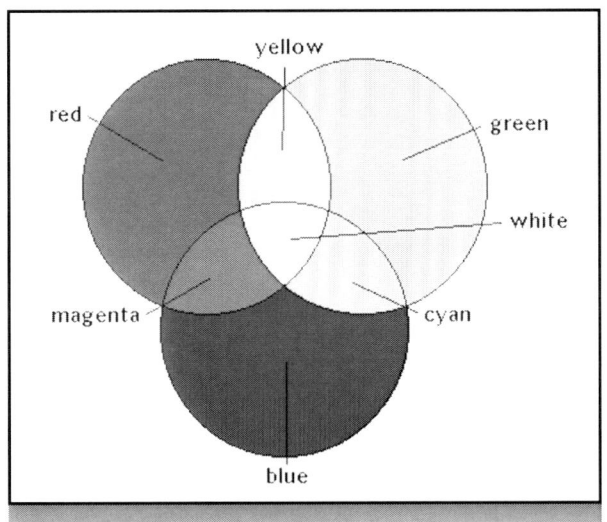

The additive colour wheel

The screen of a colour monitor, in effect, displays three images superimposed on each other. These images present, respectively, the red, green, and blue components of the colours in the scene. Colour television achieves reproduction of the wide range of natural colours by adjusting the relative brightness of these red, green, and blue images. With only these three primary colours, all possible colours and saturations can be reproduced, including black, white, and all shades of grey. This colour representation process is called the additive colour system.

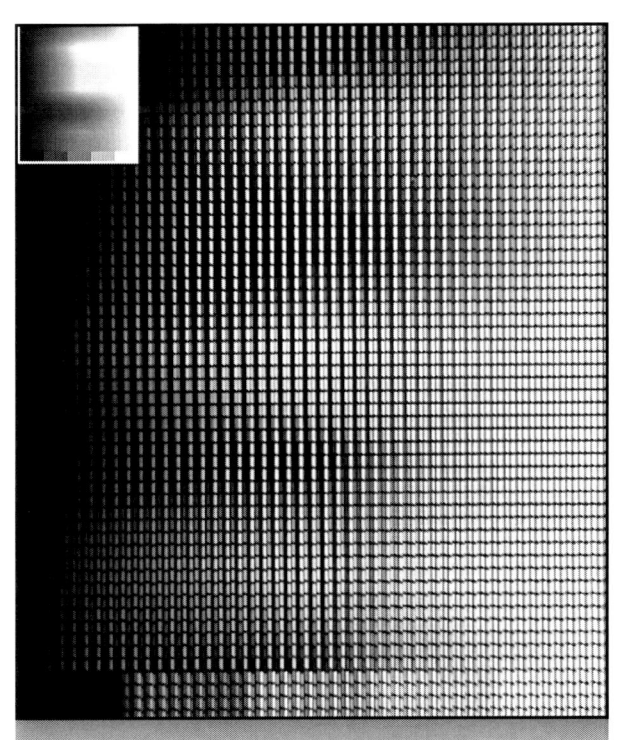

Close-up of LCD screen displaying colour palette in a computer paint program

Colour Picture Monitors

On the screen of a colour picture monitor, clusters of red, green, and blue luminescent cells are spaced very closely together. Each coloured cell responds only to the channel that it's supposed to; for example, the red cells only change their brightness level relative to the information in the red channel of the video signal at a particular position on the screen. Because the coloured dots are so small that they cannot be seen separately by the viewer, the effect of the three superimposed colours, when viewing the screen at a distance, is a colour image.

If you look at the screen of a turned-on colour monitor or television set with a magnifying glass, you can easily see the red, green, and blue cells at work. Some people have also noticed that they can see the individual picture elements if they spray the screen with window cleaner while it's turned on.

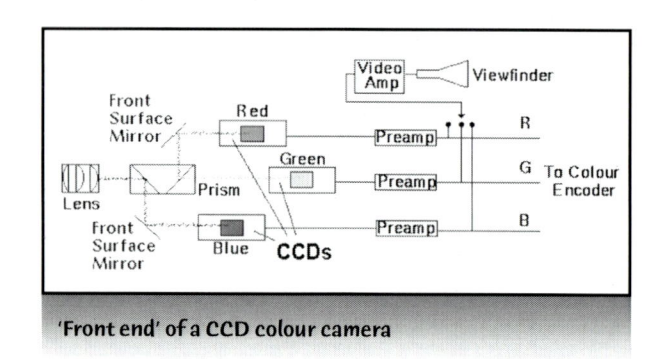

Colour Cameras

The three electrical signals that control the respective channels in the picture monitor are produced in the colour television camera by three CCD integrated circuit chips. CCD is an acronym for charge-coupled device. The camera has a single lens, behind which a set of prisms, mirrors, and coloured filters split up the full-colour scene into three different colour channels. These are focused on the three CCDs. These three electrical signals produced by the camera are transmitted to the television monitor, where the scene is re-created.

'Front end' of a CCD colour camera

Colour Encoding

- Connecting a colour video source

- Sending 3 channels colour down a sin cable.

Making luminance

- is the white & information in television created inside a colour camera.

STEP ①

Making luminance

It's created by mixing electronically three colour signals RED 30%
GREEN 59%
Blue 11%

This black-and-white signal (incorporating black-and-white information and synchronization signals) can be sent down one connection, as we discovered earlier in this chapter. How can we add colour and still keep the whole process compatible with our black-and-white system?

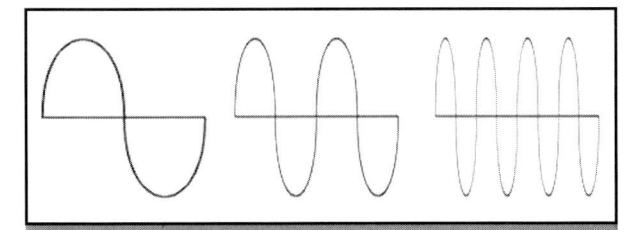

Different frequencies of a sine wave: black-and-white television is a complex mix of many different waves like these, which make up the television picture and its details

Step 2: A Special Carrier Wave

A black-and-white television signal has a full range of frequencies between about 30 Hz and 4,200,000 Hz (4.2 MHz)—that's how we transmit light and dark information and fine details within the television picture. To add colour information to this signal, we utilize a particular high frequency within that range, and use it to transmit the colour information.

The frequency used to represent colour information is exactly 3.579545 MHz, but most television engineers round off that number, and remember it as 3.58 MHz. This frequency is high enough that it won't be seen on black-and-white television sets (except as a mesh of fine detailed "dots," if we're looking closely for it). But it will still be within the bandwidth of what we're allowed to transmit over the airwaves. This special frequency is called colour subcarrier.

Colour subcarrier needs to be manipulated in some way to represent elements of colour—what the actual colour is (e.g., red, orange, yellow, green, blue, indigo, violet, and so on), and how much colour there is at a particular point in the scene (e.g., vivid or pastel, which we refer to as saturated or desaturated). We can't change the subcarrier's frequency, since frequency changes are how we tell the television system about the details in a scene. However, there are two other things we can do with it. We can change its amplitude (i.e., level), and change its phase. And that's what we do, with a device called a colour encoder.

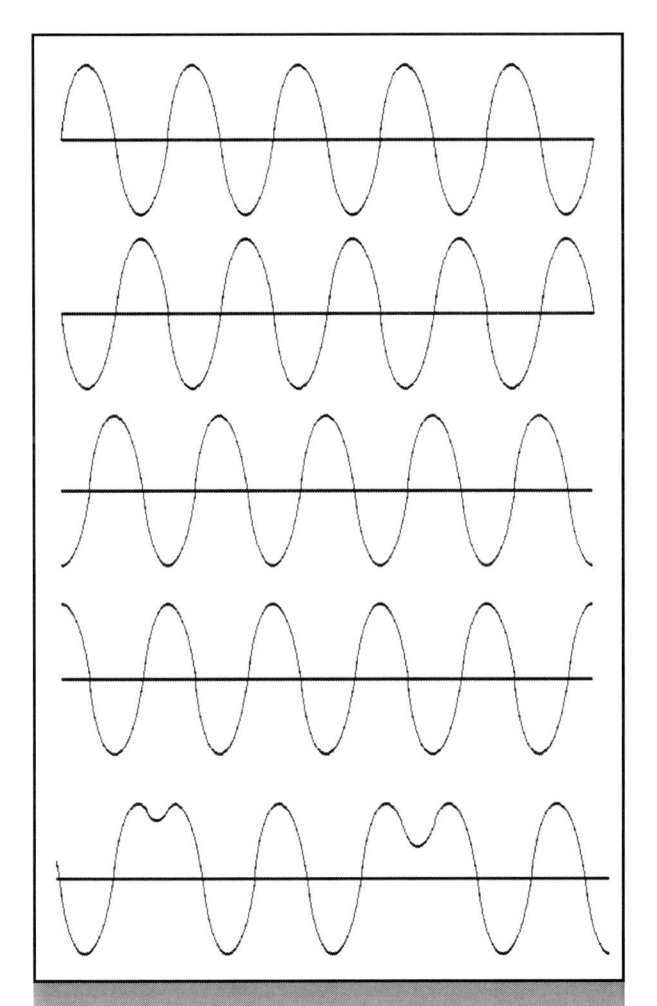

Various phases of a frequency, when mixed together, make a complex, constantly phase-changing wave. Within all those different phases are encoded all the 'hues' across a particular line of video

44 Television Technical Theory

Depending on what colour is present in the video signal at any moment in time, we send a particular phase of the subcarrier. Every possible colour has a particular phase of subcarrier. The intensity of the colour is proportional to the amplitude of the subcarrier. A high amplitude results in a saturated colour; if the subcarrier has no amplitude at all, the video signal at that moment has no colour at all—it's black, white, or a shade of grey. Through this process, we are able to produce all of the possible hues we need to, as well as information about how saturated the colours are.

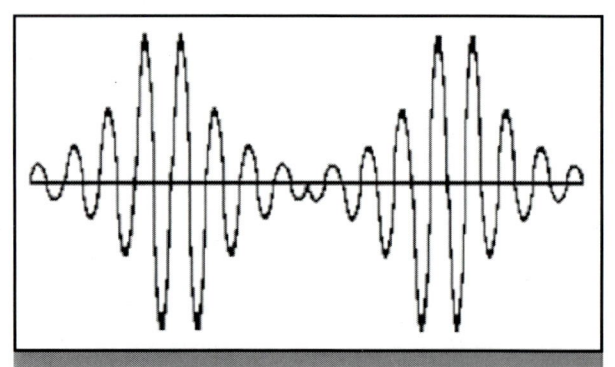

After phase modulation, the complex wave is amplitude modulated (its 'volume' is changed), representing how sat-

Step 3: A Burst of Colour Information

There is also a separate reference signal called colour burst that is added at the beginning of each video line, just after the horizontal sync signal, but still during the horizontal blanking interval. This is a short blip of colour subcarrier, and is used as a reference to give a point" with regard to which colour is supposed to be ted that colour is along the line of video. The invention system decided that colour burst will be sent at

...dients Thoroughly...

the colour subcarrier—continuously changing its amplitude (colour burst itself) is mixed with our already created so that the entire composite can be sent down a single

...vision Monitor Use ...ation Together?

...o signal, and in it you can clearly see the colour burst (within the horizontal interval), and the rest of the video ...colours within it:

Handwritten notes:

STEP ③

COLOR BURST

A SEPERATE REFERENCE signal added @ the beginning of each video line — AS A REFERENCE to what colour is going to be used and how saturated it is

STEP ④

Sending all this information done a Single Channel

---

horizontal interval, with colour burst (the lighter area)   the rest of the video line, with different colours in it (and even some pure black and white areas)

Chapter 5 Analog Video — 45

And here's how the colour encoder, inside the camera, takes the colour information from the scene, and converts it into different phases (representing the hues) and amplitudes (representing the saturation of the colour) at that particular point in the video line. The result is the complex wave at the top of the illustration. You can see that this wave can be mixed with the existing black-and-white signal, and the two signals won't interfere with one another. The television monitor is easily able to separate these two signals later on, and use the information in each one to re-create a full colour television picture:

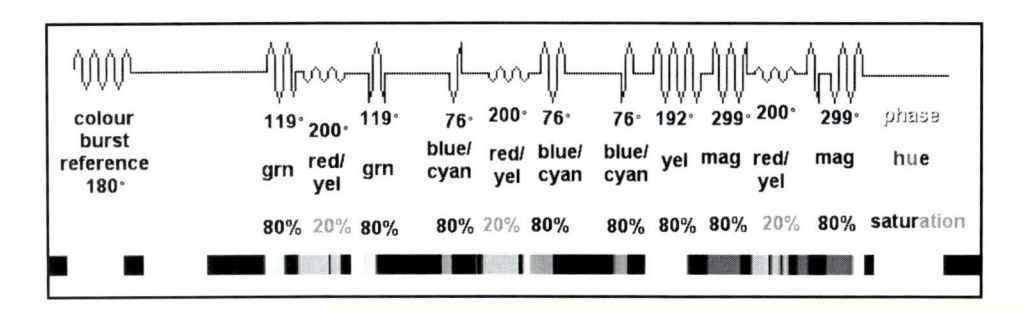

	119° 200° 119°	76° 200° 76°	76° 192° 299° 200° 299°	phase
colour burst reference 180°	grn red/yel grn	blue/cyan red/yel blue/cyan	blue/cyan yel mag red/yel mag	hue
	80% 20% 80%	80% 20% 80%	80% 80% 80% 20% 80%	saturation

Colour Video I[s]

Up until now, we've bee[n]
ond. Well, that's not exa[ctly]

In early black-and-white
video. 525 lines scanne[d]
Hz (525 × 30). With the
colour subcarrier on a bl[ack]
tle dots" referred to earl[ier]

The scanning rate of col[our]
frequency of colour subc[arrier]
ning rate for a line of co[lour]
Hz)—slightly slower than

Because the new line sc[an]
ating exactly 30 frames
29.97 frames a second
analog video only displa[ys]

ANALOG VIDEO

A Little Histor[y]

In 1936, most of the ne[w]
worked out by two comm[ittees]
committees decided th[at]

Handwritten notes:

- 525 Lines × 30 FRAMES
- Black & White TV = 30 FRAMES A SEC
- Colour TV = 29.97 FRAMES A SEC
  - ↳ scan Rate is slower & the TV wasn't able to keep up with 30 FRAMES A SECOND — it so grainy
  - ↳ SUB CARRIER

46 ■ Television Technical Theory

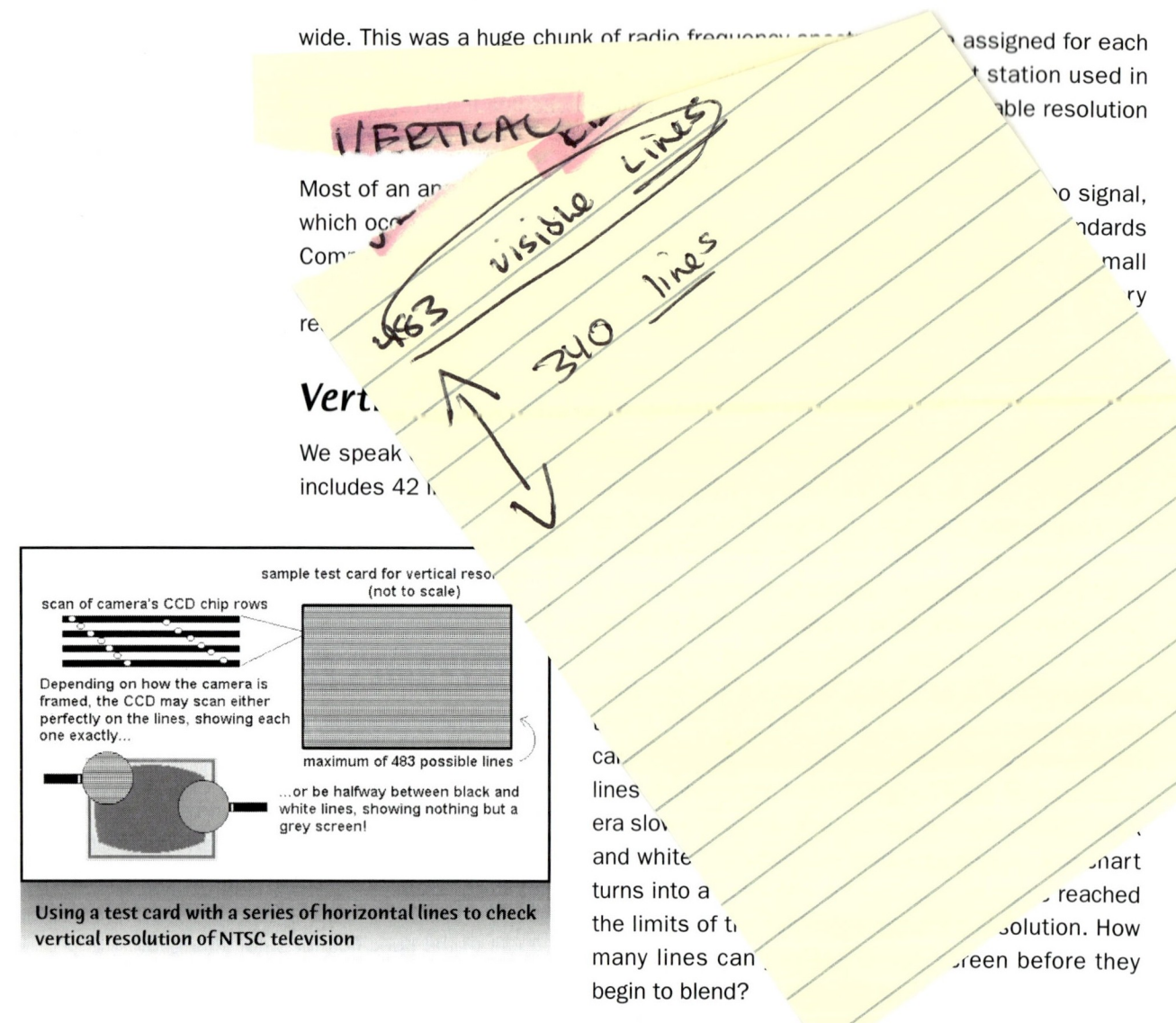

wide. This was a huge chunk of radio frequency ~~~~~~~~~ assigned for each ~~~~~~~~~~ station used in ~~~~~~~~~~ ble resolution

Most of an a~~~~~~ o signal, which oc~~~~~~ ndards Com~~~~~~ mall re~~~~~~ ry

Vert~~~~

We speak ~~~~~~
includes 42 ~~~~

[Handwritten note: VERTICAL, 463 visible lines ↕ 340 lines]

[Figure: Using a test card with a series of horizontal lines to check vertical resolution of NTSC television. Shows scan of camera's CCD chip rows, a sample test card for vertical resolution (not to scale) with maximum of 483 possible lines. Depending on how the camera is framed, the CCD may scan either perfectly on the lines, showing each one exactly... ...or be halfway between black and white lines, showing nothing but a grey screen!]

ca~~~~
lines ~~~~
era slo~~~
and white ~~~~~~~~~~~~~~~~~~ hart
turns into a ~~~~~~~~~~~~~~~~~ reached
the limits of t~~~~~~~~~~~~~ olution. How
many lines can ~~~~~~~~~~~~ reen before they
begin to blend?

If we're really careful, we might be able to scan exac~~y 483 black and white lines—one line on the chart being scanned exactly by one line of video. But, the odds of this happening are pretty dismal. And, if by some chance, we vertically reposition the chart within the camera's framing just a little bit, so our video camera's scanning lines each straddle a black and a white line, the result will be a totally grey screen with no lines visible at all! As we play around with this game of chance, it turns out that we can successfully and reliably reproduce 340 black and white lines from the chart. That's our practical vertical resolution. This fooling around, with aiming a television camera at horizontal stripes on a graphic chart, can be mathematically estimated, and it has a name. The Kell factor, as it's called (named after Raymond D. Kell), tells us that our television system resolution for interlaced analog video is about 70% of the total number of lines available to us. If we use the Kell factor in a calculation, we get $483 \times 70\% = 340$ lines of vertical resolution.

One thing confuses a lot of people about measuring resolution with this chart. They see that we have a chart with a series of horizontal lines on it, so they assume we're measuring horizontal resolution. That's not true. What we're determining is vertical resolution, and we use a chart with horizontal lines to measure it. Image a ruler placed vertically alongside the left edge of a screen. The centimetre and millimetre markings that we would use to measure the vertical height of the screen are horizontal, just as the lines in the chart are horizontal. Don't get this mixed up!

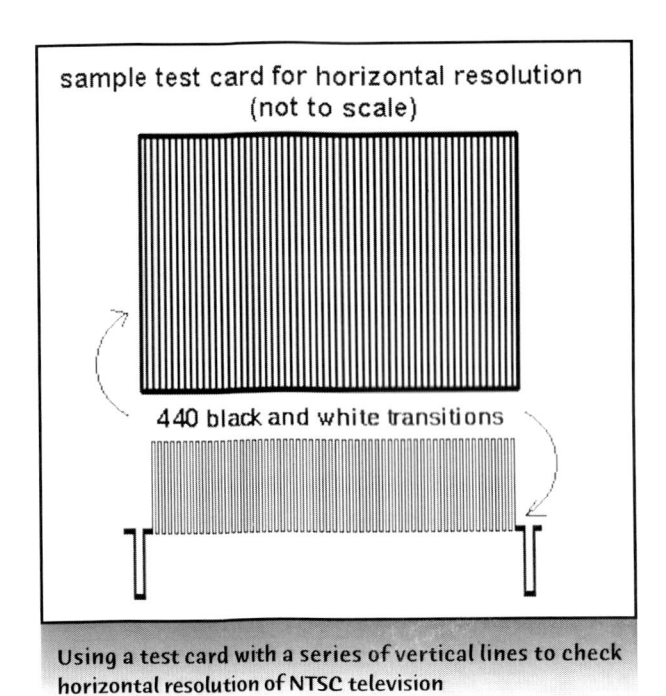

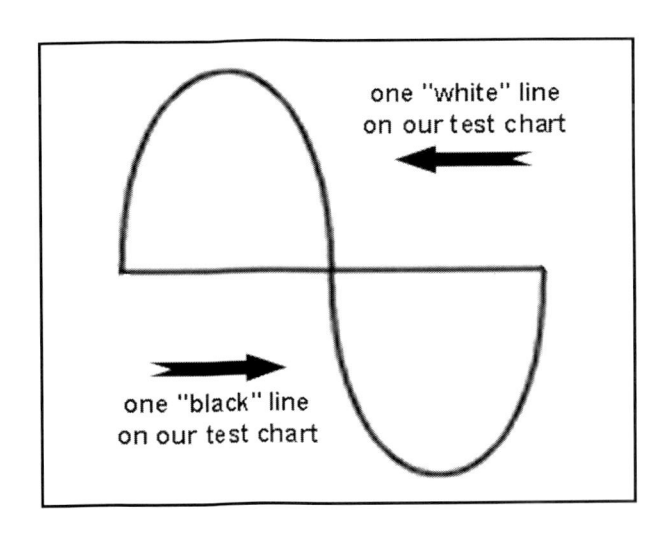

Horizontal Resolution

Horizontal resolution takes a little more math to calculate. We think of our analog television scheme as a 525-line, 29.97 frame per second television system, that takes 1/15,734 of a second for each line to scan across the screen. This is a calculation we made earlier in the chapter. But remember that some of that scan line time is used to get the electron beam back from the right side of the screen over to the left again. This is called horizontal retrace, and when we subtract that small moment of time from the total scan line time, it leaves us with a practical visible scan line that takes about 1/18,975 of a second to track across the screen from left to right. It's the visible part of a scan line that we're interested in, when we're concerned with measuring the horizontal resolution of analog television, of course.

If we take the 4.2 MHz (4,200,000 cycles per second) of bandwidth that we're allowed to use to transmit an analog television channel, and divide it by the 1/18,975 of a second that we have to display our line of video in (4,200,000 / 18,975), we get about 220 cycles of signal per line. But these are full cycles we've just calculated, not individual different brightness levels. Remember that each one of these "cycles" is a positive-going voltage followed by a negative-going voltage. We represent video by a series of ever-changing voltages corresponding to the light level read by the camera. So, in fact, each cycle would represent two pieces of video information—a lighter portion, followed by a darker portion. So with 220 cycles available to us, we could see 440 lines of video across the screen.

To consider this in practice, if we were to take our chart with black and white lines in it, and rotate it so the lines were vertical, and did the same experiment as we did with our horizontal resolution check, how many lines (black lines and white lines) would we see as we zoomed out on the image, before we began to see solid grey? Each black line would be a low voltage, followed by each white line, a high voltage. That would be one "cycle." You would find, as in our calculation, that as you pulled back from the chart, you would be able to see about 440 horizontal lines of picture information before the image was blurred into a solid grey.

To recap: the practical resolution of our analog television environment is 340 lines of vertical resolution (top to bottom), and 440 lines of horizontal resolution (left to right).

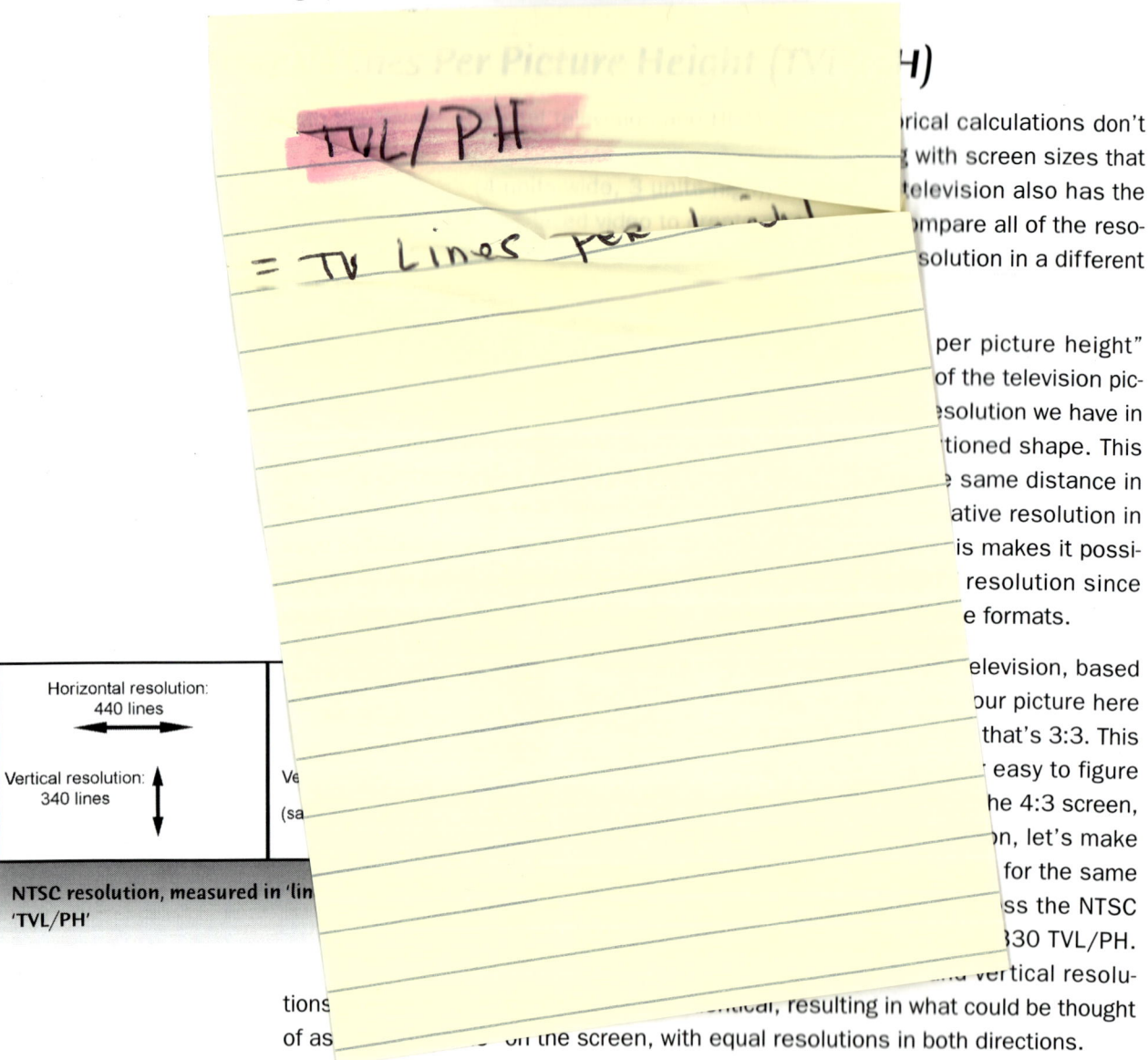

Lines Per Picture Height (TVL/PH)

[...]rical calculations don't [...] with screen sizes that [...] television also has the [...]mpare all of the reso[...] solution in a different

[...] per picture height" [...] of the television pic-[...]esolution we have in [...]tioned shape. This [...] same distance in [...]ative resolution in [...] is makes it possi-[...] resolution since [...] e formats.

[...]elevision, based [...] our picture here [...] that's 3:3. This [...] easy to figure [...] he 4:3 screen, [...]n, let's make [...] for the same [...]ss the NTSC [...]330 TVL/PH. [...] vertical resolutions [...]ical, resulting in what could be thought of as [...] on the screen, with equal resolutions in both directions.

So, the vertical resolution of NTSC analog television can also be expressed as 340 TVL/PH and the horizontal resolution as 330 TVL/PH.

NTSC resolution, measured in 'lin[es]' 'TVL/PH'

Horizontal resolution: 440 lines
Vertical resolution: 340 lines

Handwritten note: TVL/PH = TV Lines per [...]

6

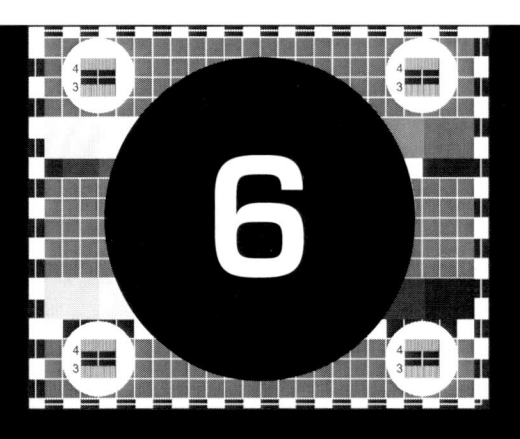

DIGITAL VIDEO

"Inside every digital circuit, there's an analog signal screaming to get out."

—Al Kovalick, Hewlett-Packard

A REVIEW OF ANALOG AND DIGITAL WORLDS

In an analog world, time is continuously observed. In a digital world, time is sampled.

An automobile speedometer is a familiar example of a device that can employ either digital or analog forms. The needle of an analog speedometer often wobbles back and forth near the printed numbers on the speedometer scale, as it responds to slight fluctuations in speed. A digital speedometer, on the other hand, displays precise numeric values for time intervals, and may even calculate an average over several time intervals to eliminate rapid small changes in the display. Although the speed is displayed in the customary decimal system, the electronic handling of values by a digital speedometer uses computer binary arithmetic.

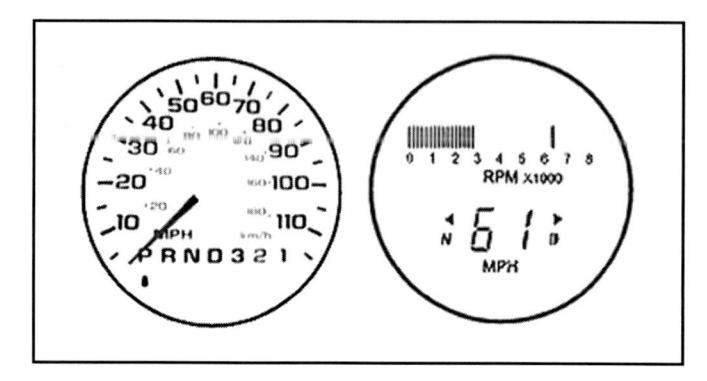

The relationship between digital and analog representations is clear when one thinks of an analog technology as producing an infinite number of samples. This implies that infinitely small amounts of time elapse between successive evaluations of an event. The digital notion follows directly from this idea by extending the time between successive samples. As the time between samples becomes large, however, the opportunity arises for the original signal to change without being sampled. Too large an interval between samples results in a less-than-accurate digital representation of the analog signal.

There are certain advantages, however, to digitizing an event, which are not available with analog monitoring. For example, it's easy to save lists of digital values, and to perform computations on such lists. This is the basis for digital signal processing. In addition, a complex event can be preserved in digital form, and analyzed or reproduced on demand, even at a speed other than that of the original event. It is also possible to process digitally encoded information that contains errors, and to correct those errors.

WHAT IS ANALOG TECHNOLOGY?

An analog device is something that uses a continuously variable physical phenomenon to describe, imitate, or reproduce another dynamic phenomenon.

A good example of this is the technology used in recording and reproducing sound on a vinyl disc. On the phonograph record, sounds are encoded in a groove that varies continuously in width and shape. When a stylus passes along the groove, that analog information is picked up and then electronically amplified to reproduce the original sounds. However, any number of minor imperfections (e.g., scratches, warps) in the record's grooves will be translated by the player into additional sounds, distortions, or noise.

WHAT IS DIGITAL TECHNOLOGY?

Digital devices employ a finite number of discrete bits of information ("on" and "off" states) to approximate continuous phenomena. Today many analog devices have been replaced by digital devices, mainly because digital instruments can better deal with the problem of unwanted information (such as noise).

In the digital technology of the compact disc for example, sounds are translated into binary code, and recorded on the disc as discrete pits in an aluminum base. Noise is less of a problem because most noise will not be encoded, and noise that does get encoded is easily recognized and eliminated during the retranslation process. A digital process has one drawback, however, in that it can't reproduce every single aspect of a continuous phenomenon. Contrary to popular belief, in a digital environment, there will always be some loss, however small. An analog device, although subject to noise problems, will produce a more complete, or truer, rendering of a continuous phenomenon.

Digital Recording and Playback in Audio

In analog recording systems, a representation of the sound wave is stored directly in the recording medium. In digital recording what is stored is a description of the sound wave, expressed as a series of "binary" (two-state) numbers that are recorded as simple on-off signals. The methods used to encode a sound wave in numeric form and accurately reconstruct it in playback were developed during the 1950s and 1960s, notably in research at the Bell Telephone Laboratories.

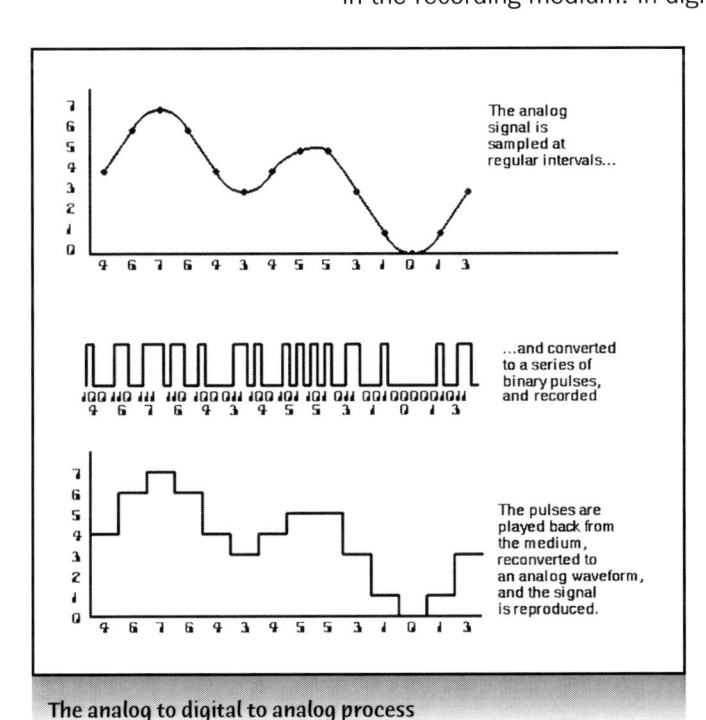

For example, when audio CDs are recorded, a "sample and hold" circuit momentarily freezes the audio waveform and holds its voltage steady, while a "quantizing" circuit selects the binary code that most closely represents the sampled voltage. This happens at regular intervals (approximately 44,000 times per second). In a 16-bit system the quantizer has 65,536 (2^{16}) possible signal values to choose from, each represented by a unique sequence of ones and zeros, 16 of them to a sample. With 88,000 16-bit conversions per second (44,000 in each

The analog to digital to analog process

The analog signal is sampled at regular intervals...

...and converted to a series of binary pulses, and recorded

The pulses are played back from the medium, reconverted to an analog waveform, and the signal is reproduced.

channel), a total of 1.4 million code bits are generated during each second of music, 84 million bits a minute, or five billion bits per hour.

Much of the circuitry in a CD player is devoted to detecting and correcting any bit reading errors that might be caused by microscopic flaws, disc pressing defects, dust, scratches, or fingerprints. Error correction is based on "parity" testing. When the recording is made, an extra bit is added at the end of every digital code, indicating whether the number of "ones" in the code is odd or even. In playback this parity calculation is repeated to detect whether any bits have changed. By cross-checking parity tests involving various combinations of the bits in each code, it is possible to identify exactly which bits are wrong, and to correct them, reconstructing the original code exactly.

This high-speed arithmetic is simple work for the microprocessor that is contained in every CD player. The data samples are also "interleaved" on the disc in a scrambled sequence, so samples that originally were one after another in time are not neighbours to each other on the disc. Correct order is restored during playback, by briefly storing the digital data in computer memory and reading it back in the proper sequence. During this de-interleaving, any large block of false data caused by a scratch or pressing flaw will be split into small groups of bad data between good samples, making it easier for the parity-checking system to identify and correct the lost data bits.

How Often and How Much?

As you can see in the "analog to digital to analog" illustration, our reproduced analog signal doesn't look quite the same as the original one—it's made up of a series of "steps." If you listened to the audio signal represented by the final waveform, it would be noticeably "raspy" and wouldn't have the high fidelity of the original. The key to making the reproduced analog signal as identical as possible to the original one is to sample often enough (sample rate) and with enough possible "steps" (high enough quantizing resolution.)

For the sample rate, the calculation is easy. You must always sample with a rate at least twice as fast as the highest frequency you want to reproduce. This is called the Nyquist theorem. In the case of audio, for example, in which we can hear frequencies up to 20 kHz, the sampling rate must be at least 40,000 times a second. In fact, CDs have a sample rate of 44.1 kHz, just to be sure that everything we can hear is reproduced. Some digital recording studios sample the original analog sources at as high a rate as 262 kHz but there is clearly some debate about whether this high a sampling rate is actually necessary, or indeed even discernable by the listener.

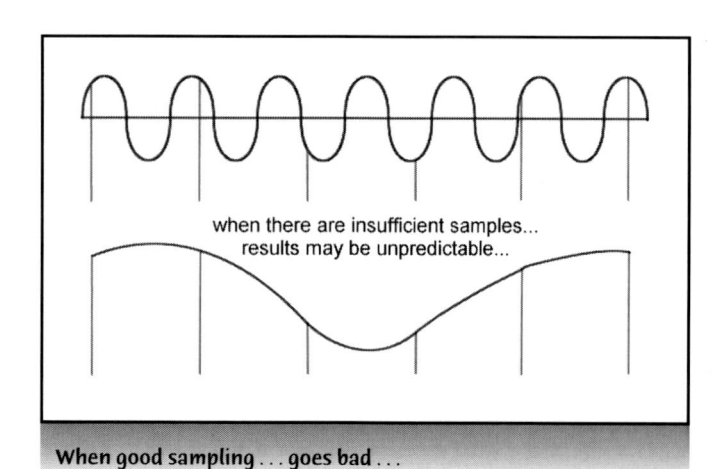

when there are insufficient samples... results may be unpredictable...

When good sampling... goes bad...

If the Nyquist rule isn't followed, and the sample rate isn't high enough for the signal we're trying to digitize, strange things happen when we try to convert the signal back to its analog form. In the accompanying picture, notice that we have sampled a wave in the first figure. But when we try to reproduce the signal based on our samples, we get a different wave altogether. This is because we haven't sampled often enough for the high frequency of the original signal.

For the quantizing resolution, more experimentation is needed. In the case of audio, research has found that 256 possible levels of voltage change (as the electrical audio wave moves up and down in level) are enough for decent audio resolution. This number—256—might seem peculiar, but because all of this digital world works with the computer binary system, it really isn't all that odd. Its other name is "eight bit resolution," and it is expressed mathematically as 2^8. This means that there are eight possible "0" and "1" combinations that could represent all the levels of change. Many people find that eight bit resolution isn't enough (they can hear the digitization, the raspy distortion,) so they prefer to sample audio at 16 bits for CD-quality sound.

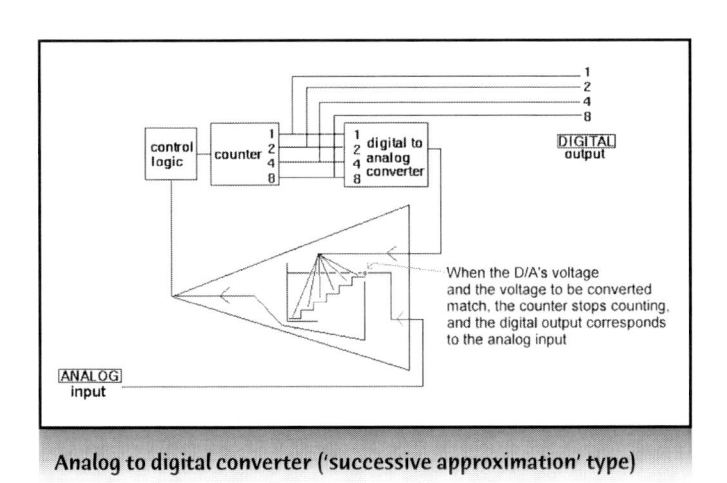

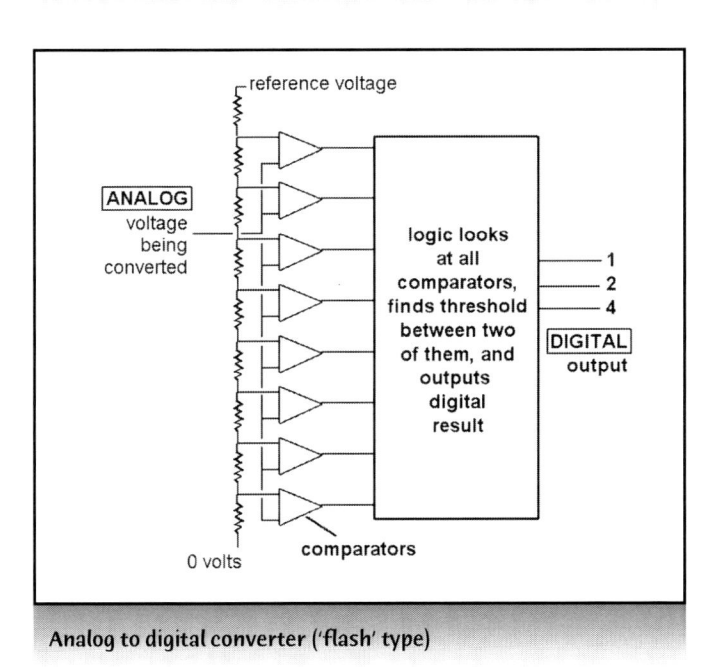

Analog to digital converter ('successive approximation' type)

Analog to digital converter ('flash' type)

Analog to Digital Converters

The basic component in an electronic digital system is the analog-to-digital converter (ADC or A/D). It converts the voltage to be measured (an analog quantity) into a discrete number of pulses (a digital quantity) that can then be counted electronically. Analog signals are electrical voltage levels; a digital computer can only handle discrete bits of information. The A/D converter thus allows a physical analog system to be processed directly by digital devices. The digital representations being converted to are in the form of binary numbers, and an ADC's precision is given by the number of binary bits it can produce as output. For example, an eight-bit ADC will produce, from its analog input, a digital output that can have 256 levels (2^8).

A typical "successive approximation" A/D converter is made from a register that can hold a digital value, an amplifier, and a voltage comparator. The register outputs are electrically summed to produce an electric current proportional to the digital value of the register. This current is amplified and compared to the unknown input analog signal. As long as there is a discernible difference, the register value changes one step at a time until there is no difference. Finally, the register holds the digital value equivalent to the analog input. This process is known as sampling. The faster the A/D converter can produce fresh samples, the more accurate will be the digital representation of the analog signal.

A "flash" A/D converter, on the other hand, has a whole bank of comparators, one for each "bit" of the binary output. The advantage of this design is that it is much faster than a successive approximation converter. However, the cost to manufacture flash converters is higher, as there are so many more voltage comparators in the design.

Digital to Analog Converters

A digital-to-analog converter, on the other hand, abbreviated as DAC or D/A, converts digital representations of numbers back into analog voltages or currents, which is what is required any time a digital image is reproduced, or a sound is played back, so that we can see or hear the program material. DACs may be constructed in several ways, the simplest of which is called a weighted resistor network. In this kind of device, each incoming bit is used to apply a voltage to the end of a resistor, which is an electronic component that blocks or resists a certain amount of electrical current. If the bit is "1," the switch is turned on for that particular bit, and the voltage is high; otherwise the voltage is low. The resistances at the outputs of these switches vary as powers of two (each electrical resistance is double the previous one on the network), with the least significant bit of the binary output being applied to the largest resistance. So, the maximum current flow into each resistor is proportional to the binary weight of the bit, and the total current flowing out of all of the resistors combined is proportional to the binary value of the input. The switches are very fast, so we can create a continuous analog output as quickly as we can input different digital values.

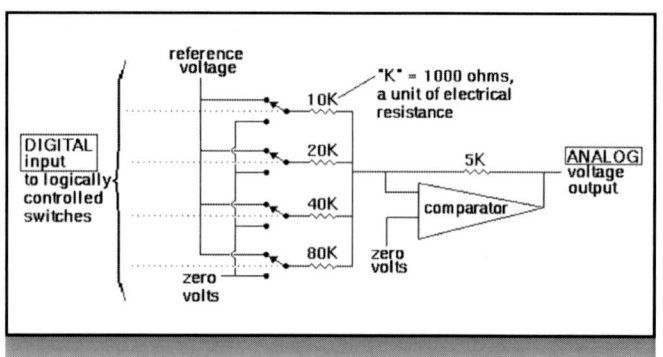

Digital to analog converter ('resistive ladder' type)

DIGITAL VIDEO

Why should we convert an analog signal (viewed with our "analog" human eyes) into a series of "1"s and "0"s? There are several reasons. Digital transmission and processing generates a minimum of degradation (if the signal is recovered properly)—a zero is a zero, and a one is a one, providing they're received anywhere within the voltage range of where they should be. As a result, there is no problem with noise and other deterioration as with analog methods. Interfacing with computers and video manipulation devices (e.g., digital video effects devices (DVEs), stillstores, frame synchronizers, graphics systems) becomes much easier if no conversions from analog are required before this processing takes place.

But First, A Few Words from Our Analog Video World . . .

In legacy analog videotape libraries, there will generally be two types of recordings: composite and component.

Composite Recordings

Composite recording takes the full NTSC video signal (with its chrominance "lay-
e … se of many pro-
f … l modulates a
h … the videotape.
S … vith filters, and
t … ple, in the case
c … recorded with a
r … why we refer to
s … g recording suf-
f … nce parts of the
p

Composite Recordings

Luminance is Recorded with high noise levels = ANALOG high frequency levels

Component Recording

LD ~~uses R~~

R, G, B is mixed together to create -Y

then ran through a mixer to create

R-Y, B-Y colour component information.

(

… blem associated
… nt signals (called
… her. In the NTSC
… , G, and B infor-
… inance (Y) infor-
… s then inverted
… R, B, and –Y are
… form R-Y and B-Y
… . In component
… colour encoding
… channels are
… oint. There is no
… olour subcarrier,
… on. Legacy com-
… uites were com-
… cial component
… to maintain the
… leo over compos-
ite video.

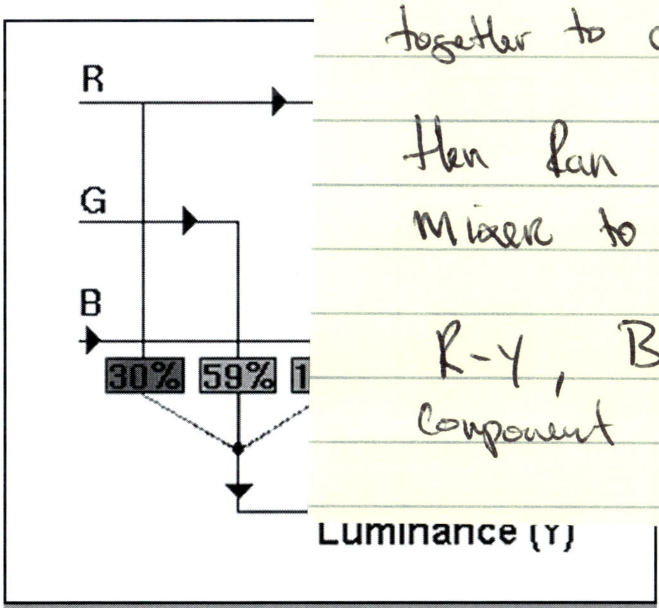

Component video

Standard Definition Video

How Many Samples Does It Take?

Composite Video

When the digitizing of video first began, most of the time we were dealing with composite video from analog videotape machines. It was decided to sample this video at 14.31818 MHz, which is four times the frequency of colour subcarrier. If we do the math, this gives us 910 samples for the entire line of video, and 768 of these are for active video. In those early days, we sampled composite video at either a 10-bit or 8-bit quantization. As some of those bits were reserved for other technical purposes, there were only 1016 discrete levels of video in the 10-bit system, and only 254 levels in the 8-bit system. Also, when digitizing composite video with this method, all of the video was digitized, including the blanking, sync signals, and colour burst.

In the professional video world, new video content in the analog composite video format has not been produced for some years, so this original digitizing format is no longer used. Today, even if a legacy analog videotape is to be digitized from the archives, it is converted to component video by the digitizing hardware first, and then sampled using the component video method, described next.

Component Video

In the component world, we have Y (luminance), B-Y and R-Y. We sample the luminance signal at 13.5 MHz. This frequency was chosen because it affords a certain compatibility with NTSC (North American) and PAL (European) digital video streams—yes, actual international agreement on a television standard. If we allow for the fact that the horizontal interval takes a certain amount of time to occur within each video line, this results in 720 luminance samples for a line of active video – the portion of the video signal that we actually see. We can sample component video at either a 10-bit or 8-bit quantization, but 10-bit is now more common. For the luminance channel, there will be only 877 discrete levels of video in the 10-bit system (you might expect 1024 levels, or 2^{10}), and only 220 levels in the 8-bit system (instead of 256, the full 2^8), as the remaining levels are reserved as a safety margin.

The B-Y and R-Y information is sampled at only half this rate—6.75 MHz—because our eyes can't really discern colour detail much more closely than that. Researchers learned this fact while developing the NTSC chrominance channels back in the 1950s. For each chrominance component, therefore, the sampling is half as often: 360 samples for each line of active video. For the each of the chrominance components, there will be only 897 discrete levels of video in the 10-bit system (not 1024), and only 225 levels in the 8-bit system (not 256). You will notice that there are roughly the same number of quantizing levels in both the chrominance and luminance channels, even though the sampling rate of the chrominance information is only half that of the luminance.

The sampling rates of component digital video can also be looked at another way. For every four (4) samples of luminance, there are two (2) samples of R-Y, and two (2) samples of B-Y. This is expressed in a shorthand way, and standard component digital video is often called "4:2:2" video.

Component serial video has special timing reference signals (TRS), and they are inserted after the leading edge of the horizontal sync pulses, during the conversion from analog to digital. These signals indicate, within the serial bit stream, the beginning and the end of the video lines and fields. As well, ancillary data can be sent in lines 10–19 (and 273–282 in the second field), as well as vertical sync, horizontal sync, and equalizing pulse information. Audio information in this data segment has been standardized as SMPTE 272M, and this means that audio (up to four channels of it) can be sent with the video, down one serial video coaxial cable.

Let's Start Sending Those Digital Video Signals

So, we've figured out how to convert analog video into a digital format, and send that stream of information down a coaxial cable.

A component stream is a series of three interleaved signals, sent in the following manner:

B-Y	Y	R-Y	Y	B-Y	Y	R-Y	Y

And so on. You will notice that we always start the stream with a B-Y colour component, then a luminance component, then an R-Y, and finally, another luminance. This sequence then repeats.

Let's see just how much information we're sending down a coaxial video cable. We have 13.5 million samples of luminance, 6.75 million samples of R-Y information, and another 6.75 million samples of B-Y information. If we were to send this, multiplexed, down a parallel transmission path, we would be sending 13.5 + 6.75 + 6.75 = 27 million samples per second.

But, we're sending this data stream down a serial cable. Each sample has 10 individual bits of information (a "1" or "0"), so our actual transmission rate down a serial coaxial cable will be 270 million bits per second.

Sampling Rates

We are now familiar with what 4:2:2 sampling is. What are the other ones we hear about? Here's a synopsis of the digital video formats:

- ■ **4:2:2**—a component system. Four (4) samples of luminance associated with 2 samples of R-Y, and 2 samples of B-Y. The luminance sampling rate is 13.5 MHz; colour component rates are 6.75 MHz. This is the highest resolution for studio component standard definition (SD) video, and is also used in regular high definition (HD) video.

Various digital video sampling rates

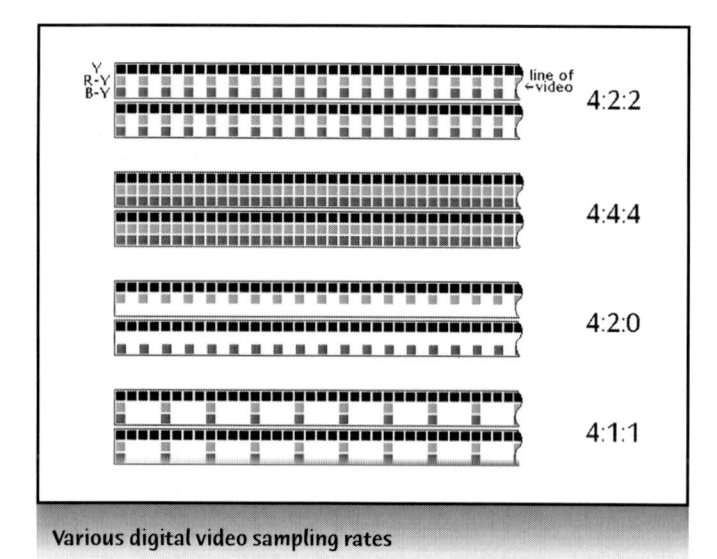

- **4:4:4**—all three components are sampled at 13.5 MHz. This sampling format is used where there will be a high degree of postproduction manipulation and compositing. The higher chroma resolution will ensure that there is no fringing in the images or other degradation in the final image. However, this sampling rate requires more storage during recording and also in the digital compositing system where the effects are produced.

- **4:2:0**—this is like 4:2:2, but while doing what's called "vertically subsampled chroma." This means that while the luminance sampling rate is 13.5 MHz, and each component is still sampled at 6.75 MHz, only every other line is sampled for chrominance information. The missing information on alternate lines is copied from the previous video line and inserted. Some consumer camcorders use this sampling format, to save space on the videotape, hard drive, or solid-state recording medium.

- **4:1:1**—the luminance sample rate here is still 13.5 MHz, but the chrominance sample rate has dropped to 3.375 MHz. It is a format commonly used in ENG (electronic news gathering) and other noncritical applications where high chrominance detail is not required.

IT'S STILL VIDEO

Synchronization Signals

There are a few more things we should mention about this serial transmission stream. Through this discussion of sampling video and turning it into a digital form, we've lost sight of the fact that it's video, and has a certain line length, synchronization pulses, and so forth, which we need to recover, or at least indicate the presence of, if we are to put this data back together as television pictures that we can view properly.

As we know, there is some "lost time" during the horizontal and vertical blanking intervals, when we're not doing much of anything useful except synchronizing our lines and fields, respectively. We can see these areas easily on either a waveform monitor, or by selecting the H and V delays on a professional video monitor.

When the new digital video format was being designed, engineers realized that this period of time could be put to better use by simply describing the H and V syncs as a short series of "sync words."

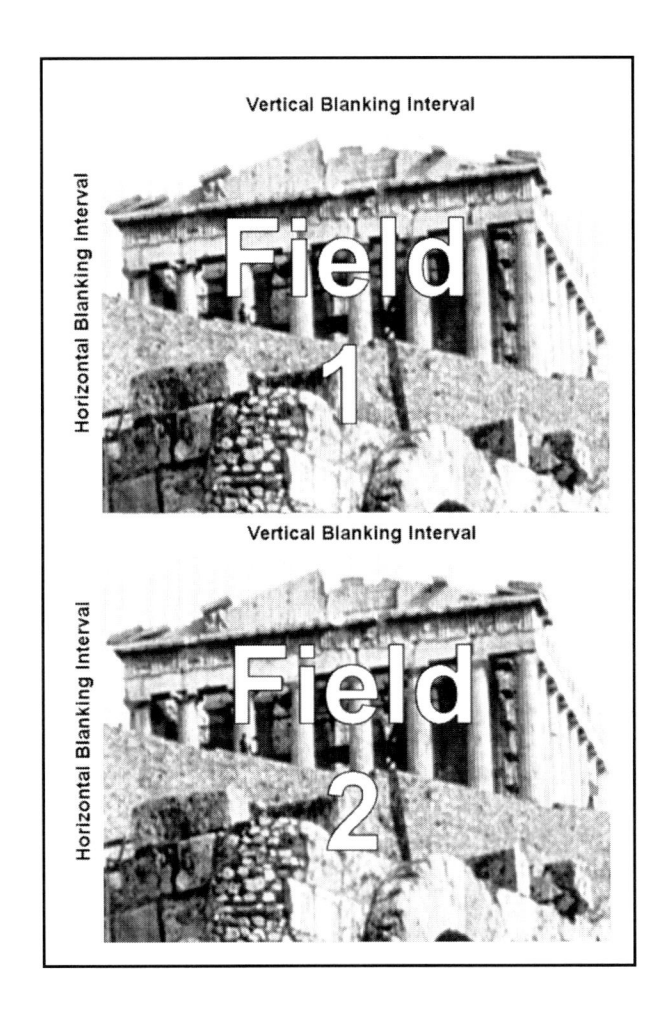

Vertical Blanking Interval

Field 1

Vertical Blanking Interval

Field 2

H and V delayed video signal on a monitor, showing the horizontal and vertical intervals

Each line of digitized video, therefore, has two more parts: special time reference signals (TRS), which are called start of active video (SAV) and end of active video (EAV). The SAV and EAV "words" have a distinctive series of bytes within them (so they won't be mistaken for video information), and a "sync word" which identifies the video field (odd or even), and presence of the vertical blanking interval and horizontal blanking interval. The TRS signals are the digital equivalent of our analog sync and blanking. We'll describe them in more detail later as we analyze the various digital video standards.

Scrambling To Get It Together

Next, all of the conversions to "1"s and "0"s may have one real problem: it's possible that we may have a long string of all "1"s or all "0"s in a row. This is not a small dilemma since, in serial transmission, we want to have a lot of transitions so that we can recover the "clock," which is the constant time reference to tell digital equipment where to find each of the ones and zeros being received through the video cable.

it's very difficult to recover this clock signal

from this NRZ (non-return to zero) serial stream, since there aren't enough transitions in it

however, if we convert the NRZ serial stream

to NRZI

there are more transitions, making the clock recovery easier

Why we need NRZI

The solution to this problem lies in something called scrambled NRZI ("non-return to zero - inverted") code. The "non-return to zero" part means that NRZI is derived by converting all the "1"s to a transition, and all "0"s to a non-transition.

The neat thing about this code is that it is polarity independent. This means that it's no longer a case of a "1" being a high voltage, and a "0" being a low voltage, but rather a transition indicates a "1," regardless of which way the transition is going. This means that the clock can be recovered more easily, since there are now more transitions in the digital signal, and there's no confusion about what a "0" or "1" is, either.

HIGH DEFINITION VIDEO

How Many Samples Does It Take?

In high definition (HD) video signals, the horizontal scanning rates are faster to accommodate many more scan lines within a field or a frame. The vertical retrace rates depend on the HD format (e.g., 720 or 1080 lines). There is much more data content, which requires faster sampling and faster transmission speeds (up to 1.485 Gbps). The quantizing level is the same as standard definition video—10 bits, meaning the HD digital video signal has a transmission rate of 1.485 GHz. The aspect ratio, of course, is 16:9 instead of 4:3. Both interlaced and progressive scans are permitted.

The TRS concept is the same as in standard definition (for horizontal and vertical blanking periods), but high definition digital video also contains line number and error detection checksum words on every line. The period between EAV and SAV is not used by normal HD video and may be used for embedded audio or time-code data, in the same way that it is used in standard definition (SD) digital video.

HDTV digital component video is produced by applying a 4:2:2 sampling structure to the analog signal. The luminance component (Y) is sampled at 74.25 MHz, and the colour difference components, U & V, (similar to R-Y and B-Y) are both sampled at 37.125 MHz. The Y stream is quantized to 10 bits resolution and Timing Reference Signals (TRS) are added at the beginning and end of the horizontal video blanking period. The U & V streams are also quantized to 10 bits and then interleaved to give a chroma stream at 74.25 MHz. The 74.25 MHz Y and C streams are then interleaved to produce a single stream at 148.5 MHz. The data is scrambled and then serialized using a Non-Return to Zero (NRZ) code to produce a 1.485 GHz signal.

Digital Video Resolution

In the Analog Video chapter, we discussed the resolution of our 1950s-era NTSC television signal. Expressed as "TV lines per picture height", it came out to about 330 TVL/PH. Notice that this is limited by the NTSC transmission system. Our broadcast-quality equipment (cameras, character generators, etc.) may have much higher resolution when we make our pictures in the studio. It's just that this higher resolution doesn't make it to the home viewer.

Digital video's horizontal resolution is determined by the sampling frequency. A physical reality called the Nyquist theorem states that the highest frequency you can reproduce is one which is one-half the sampling rate. Let's look at composite and component video to see how the resolutions come out.

Composite Digital Video

Composite digital video's sampling rate is 14.3 MHz. The maximum frequency that can be reproduced is half of this, or 7.15 MHz. Our scanning line is still 52.86 µs. long. But we can put black and white vertical stripes in each cycle of our maximum frequency. Finally we multiply the whole thing by .75 (we're dealing with a 4:3 format) to give us our new comparative number TVL/PH. The math goes like this:

   52.86 µs × 2 × (14.3 MHz/2) × .75 = 567 TVL/PH for composite digital video

If you sample in composite (14.3 MHz), you'll get 567 TVL/PH.

Component Digital Video

The formula works the same way, except that component's luminance sampling rate is 13.5 MHz, so you get a little less resolution:

   52.86 µs × 2 × (13.5 MHz/2) × .75 = 535 TVL/PH for component digital video

If you sample in component (13.5 MHz), you'll get 535 TVL/PH for a 4:3 aspect ratio system.

HDTV Analog Resolution

A longer scanning line (25.86 µs of visible line time) and a higher video line count (1080 visible lines) will result in a higher resolution. In analog HDTV systems (yes, there are HDTV systems that work in an analog system), the bandwidth limit is 30 MHz (which is much higher than our old NTSC transmitter limit). This is largely due to the fact that there are practically no HDTV analog transmitters out there and the ones that exist were designed to operate on special, non-NTSC channel allocations.

Also, we're now dealing with a 16:9 format so our "make it square" factor is no longer 3/4 (or .75) but is now 9/16 (.5625). The formula works out like this:

   25.86 µs × 2 × 30 MHz × .5625 = 873 TVL/PH.

HDTV Digital Resolution

The sampling frequency for HDTV digital is 74.25 MHz. The maximum frequency we can reproduce is half of this, because of the Nyquist theorem. The resolution is limited only by the 30 MHz bandwidth we mentioned in the HDTV analog system:

   25.86 µs × 2 × (74.25 MHz/2) × .5625 = 1080 TVL/PH

Here's a synopsis of what we've learned:

- NTSC (transmitted): 330 TVL/PH
- Digital Composite: 567 TVL/PH
- Digital Component: 535 TVL/PH
- HDTV Analog: 873 TVL/PH
- HDTV Digital: 1080 TVL/PH

HD Format Variations

Because HD incorporates different field rates, frame rates, and scanning resolution, several video standards have been approved to be used with serial digital video. These are defined in SMPTE 274M for 1080-line scanning and SMPTE 296M for 720-line scanning.

SMPTE 274M

Standard	Samples / Line	Active Lines / Frame	Scanning	Sample Frequency	Total Samples / Line	Total Lines / Frame
1920 x 1080/60/2:1	1920	1080	30 Hz interlaced	74.25 MHz	2200	1125
1920 x 1080/59.94/2:1 *	1920	1080	29.97 Hz interlaced	74.176 MHz	2200	1125
1920 x 1080/50/2:1	1920	1080	25 Hz progressive	74.25 MHz	2640	1125
1920 x 1080/30/1:1	1920	1080	30 Hz progressive	74.25 MHz	2200	1125
1920 x 1080/29.97/1:1 *	1920	1080	29.97 Hz interlaced	74.176 MHz	2200	1125
1920 x 1080/25/1:1	1920	1080	25 Hz progressive	74.25 MHz	2640	1125
1920 x 1080/24/1:1	1920	1080	24 Hz progressive	74.25 MHz	2750	1125
1920 x 1080/23.98/1:1	1920	1080	23.98 Hz progressive	74.176 MHz	2750	1125

* This standard gives an exact frame rate compatibility with NTSC.

Several sub-standards for this are defined:

SMPTE 296M

Standard	Samples / Line	Active Lines / Frame	Scanning	Sample Frequency	Total Samples / Line	Total Lines / Frame
1280 x 720/60/1:1	1280	720	60 Hz progressive	74.25 MHz	1650	750
1280 x 720/59.94/1:1	1280	720	59.94 progressive	74.176 MHz	1650	750

Some Problems with Digital Video

We've digitized the video, figured out the encoding, and we've sent the signal down the coaxial cable. It should be perfect from here on, forever.

Or is it?

With analog video, you can send the signal a long way down a coaxial cable. If the signal starts to degrade, it loses colour and detail information, but this can be boosted somewhat with an equalizing distribution amplifier, so the signal doesn't look quite as bad as it really is.

The problem with digital video is that the signal is excellent until you send it too far, at which point the signal is nonexistent! The fall-off doesn't happen over a long distance; it all plunges downhill over a critical a dozen or so metres of cable. This is called the "digital cliff" effect.

We could probably live with one minor video disturbance a minute, sending our digital video 375 metres down the cable. We will get a glitch every second if we extend that cable only 10 more metres. We will get interference 30 times a second if we lengthen the cable another 10 metres. At the critical length, a little more extension greatly increases the chances of fouling the signal. The solution to this problem is to capture the signal and regenerate it with a video distribution device, every 250 metres or so.

Compression

We can be content with sending our very high-speed digital video signals down a coaxial cable. A lot of information is going down that wire, however. This requires a fairly wide bandwidth transmission path, which is something that we do not always have available to us. Consider that the local cable company tries to push hundreds of television signals down one coaxial cable (with varying degrees of success, admittedly). Even our standard analog over-the-air television channel (with its 6 MHz bandwidth) can only handle about 25 Mbps. Most computer network interfaces have a capacity somewhere between 100 Mbps and 1Gbps. None of these transmission technologies is even near what we need for a full bandwidth digital transmission.

Let's look at this another way. A clean, uncompressed 4:2:2 SD digital component signal would need 97 gigabytes of hard drive storage for a one hour show. To facilitate storage-economical editing and postproduction, we need to reduce the bit rate.

Lossless or Lossy?

Lossless compression . . . loses no data. This allows only small amounts of bit-rate reduction to a maximum of roughly 3:1—the video information can only be compressed to a third of its original size.

Lossy compression, on the other hand . . . loses information. If we're careful with our use of the compression scheme's mathematical formulas, we can make this loss almost invisible, but it isn't perfect. If we're not careful, things become quite messy, and the result usually appears as small blocks within the video frame, or information that doesn't update properly from one frame of video to the next. However, with lossy compression, we are able to send the video signal down a lower bandwidth channel.

Spatial Redundancy

Blue Pixel Value = 118	Blue Pixel Value = 118	Blue Pixel Value = 118	Blue Pixel Value = 118	Blue Pixel Value = 118

We can reduce the signal by looking closely at it. In any given frame of television, there is probably a lot of redundant information—large areas of a single colour, for example. If we can somehow encode the information so that, when we look at a blue sky, instead of saying:

>BLUE PIXEL, BLUE PIXEL, BLUE PIXEL, BLUE PIXEL, BLUE PIXEL

we say

>FIVE BLUE PIXELS

we can save a lot of transmission time and information repetition.

Temporal Redundancy

Also, if we look at frames of video shot before and after the frame we're encoding, we might find that it's a static shot with no movement. In that case, we can save a lot of information by, instead of saying:

>A VERY DETAILED FRAME OF VIDEO
>
>A VERY DETAILED FRAME OF VIDEO
>
>A VERY DETAILED FRAME OF VIDEO
>
>A VERY DETAILED FRAME OF VIDEO

we can say

>A VERY DETAILED FRAME OF VIDEO

IDENTICAL FRAME

IDENTICAL FRAME

IDENTICAL FRAME

Block Level Compression

We can also apply a technique that allows us to use a lower number of bits for information that occurs frequently (similar to the spatial redundancy theory described above, but instead of working on the video-line or video-frame level, it's done on a "small block of pixels" level.) This allows more difficult scenes to have more "headroom" in the bit stream as it moves along, so the scenes won't be distorted.

A combination of all three methods can, with judicious use, reduce our 270 Mbps stream down to as low as 6 Mbps. There is some debate about whether this will be noticeable to the home viewer.

Basic Compression Types

- **JPEG:** This stands for Joint Photographic Experts Group and is a method for compressing still, full-colour images.

- **Motion JPEG:** a variation on the above, which is essentially a way to play back a series of JPEG images back-to-back—sort of like a "flip book" animation, where each frame of video is compressed, but there is no acknowledgment of any similarities between these individual frames.

- **MPEG:** This is an acronym for Moving Pictures Experts Group, and is an international standard for moving pictures around. There are various versions of MPEG.

- **MPEG-1:** This was the first MPEG standard (released in November 1991), and is used with CD-ROMs for moving picture playback. This is referred to as the VCD format, with a frame size of 360 x 240 pixels. It has a VHS-like quality.

- **"MPEG-1.5":** This is, strictly speaking, not really a standard. This is the application of MPEG-1 to full-size, interlaced, broadcast video. And it shows. There are noticeable "blocky" artifacts in the video (sometimes disparagingly referred to as "pixel-vision" or "chiclet-vision").

MPEG-2 Profiles and Levels							
Level	Profiles						
	Profile	Simple	Main	SNR	Spatial	High	4:2:2 Profile
	Frame Type	I, P	I, P, B	I, P, B	I, P, B	I, P, B	I, P, B
	Sampling	4:2:0	4:2:0	4:2:0	4:2:0	4:2:2	4:2:2
	Scalability	non-scalable	non-scalable	SNR scalable	spatially scalable	SNR, spatially scalable	SNR, spatially scalable
High (e.g., HDTV)	Enhanced		1920 x 1152 60 fps 80 Mbps			1920 x 1152 60 fps 100 Mbps	
	Lower					960 x 576 30 fps	
High-1440	Enhanced		1440 x 1152 60 fps 60 Mbps		1440 x 1152 60 fps 60 Mbps	1440 x 1152 60 fps 80 Mbps	
	Lower				720 x 576 30 fps	720 x 576 30 fps	
Main (e.g., conventional TV)	Enhanced	720 x 576 30 fps 15 Mbps	720 x 576 30 fps 15 Mbps (conventional television)	720 x 576 30 fps 15 Mbps		720 x 576 30 fps 20 Mbps	720 x 608 30 fps 50 Mbps
	Lower					352 x 288 30 fps	
Low (e.g., computer compressed)	Enhanced		352 x 288 30 fps 4 Mbps	352 x 288 30 fps 4 Mbps			
	Lower						

- **MPEG-2:** This specification was issued in November 1994, and is the full-motion, broadcast video version of MPEG. All broadcasters have adopted it for their compression method. MPEG-2 has a wide range of applications, bit rates, resolutions, qualities, and services. This makes it very flexible, but how do you tell the decoder device what format you have selected? We have built into the MPEG-2 format a number of "subsets" specified by a "Profile" (the kind of compression tools you're using) and a "Level" (how complex your processing actually is). MPEG-2 supports transport stream delivery (streaming video), standard definition (SD) and high definition (HD) video, and 5.1 surround sound.

- **MPEG-3:** This was to be the next step, but its variations ended up being incorporated into the MPEG-2 standard, so this version was never released by the Group.

- **MPEG Audio Layer III (MP3):** This is a format for compressing audio at extremely low bit rates. Its most common application is on the Internet for the transferral of audio, and of course for use in the now-ubiquitous MP3 player.

- **MPEG-4 Part 2:** Released in 1998, this is a standard to allow reliable video compression at lower speeds of transmission (5 Kb/s to 64 Kb/s). This means the possibility of video transmission over telephone lines and to cell phones, and high quality audio over the Internet. The starting point for full development of MPEG-4 is QuickTime, an existing format developed by Apple Computer, but the resulting file size with MPEG-4 is smaller. Other derivations of MPEG-4 include DivX and Xvid.

- **MPEG-4 Part 10:** This is more commonly referred to as H.264 or AVC, and has been adopted in a number of professional video cameras, and in other products including PlayStation Portable, Xbox 360, iPod, and Blu-ray DVD players.

- **MPEG-7:** A "multimedia content description interface." Not necessarily a compression method, per se, but more a standard for descriptive content within compression. This type of information (above and beyond audio and video information) is called metadata, and will be associated with the content itself, as textual information, to allow fast and efficient searching for particular video content or other material that you're looking for within a content management system library database.

How MPEG Works

Given that there are spatial (adjacent pixels within a frame are similar) and temporal (the same pixel between sequential frames are similar) redundancies in video, we can do compression.

In MPEG, a small block of pixels (8 x 8) is transformed into a block of frequency coefficients. What this means is that, in any given block of pixels, there will be some pixels that aren't that far off in video level from their neighbours. This is something that we can exploit in our compression scheme. What we do is perform

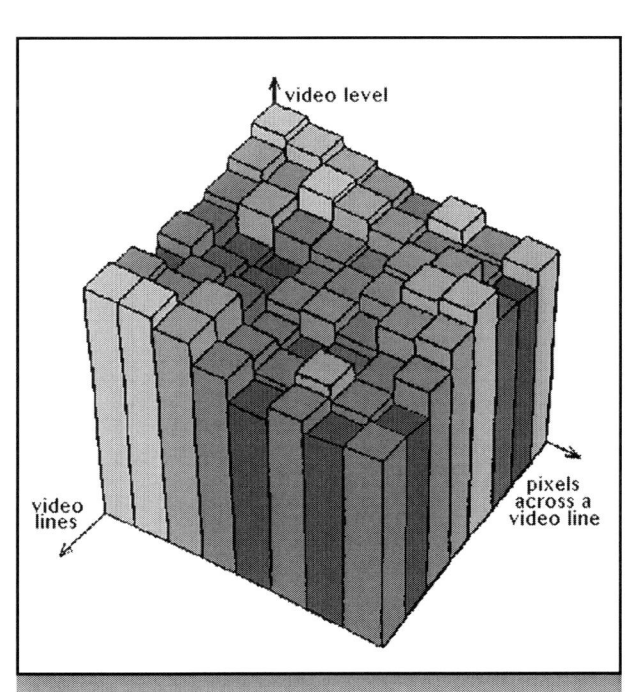

BEFORE: a graphical representation of a block of digitized pixels

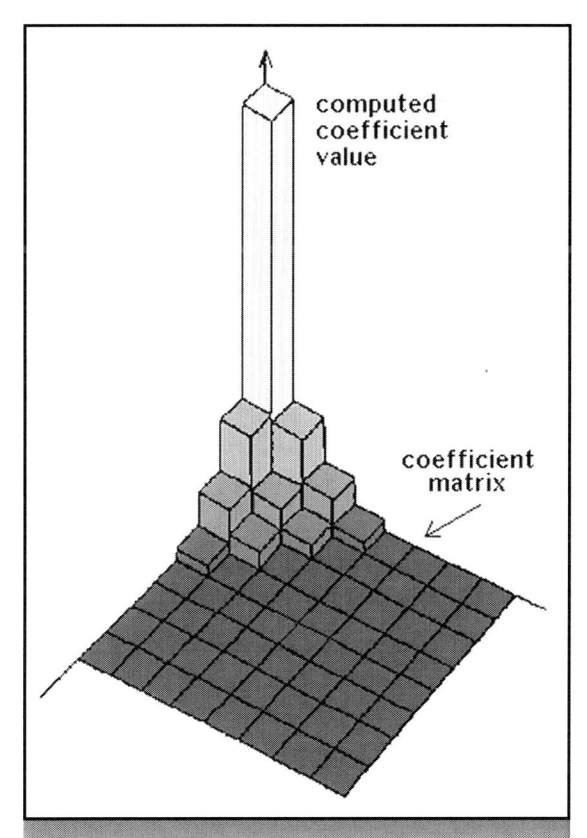

AFTER: how Discrete Cosine Transform reduces the amount of information about the block of digitized video

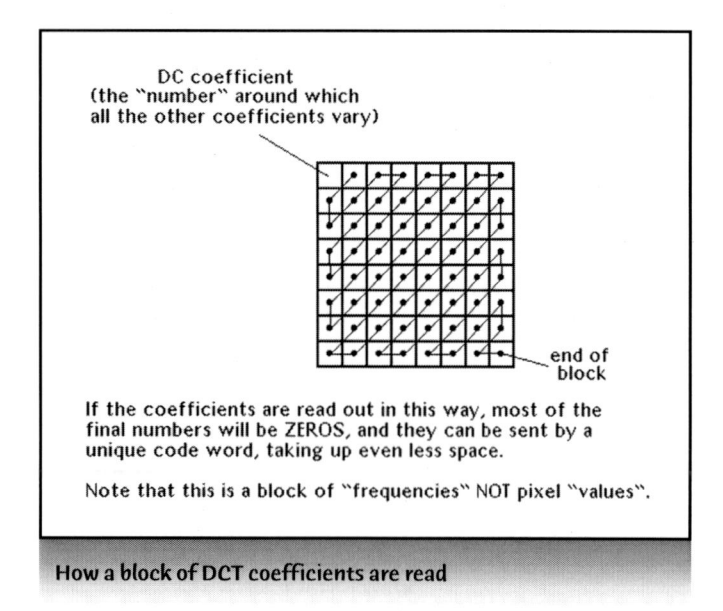

a Discrete Cosine Transform (DCT) on the block of pixels.

Without getting into the math too much, we end up with an 8 × 8 block of information representing how large the changes are over the entire block of pixels, and how many different changes there are. These are what are called frequency coefficients.

Because the block only has 64 pixels in it, and they're all right next to each other, and their values are probably pretty close to one another, most of the 64 "value spaces" available for the direct cosine transform calculation have values of zero! So, instead of sending 64 values for 64 pixels, we may end up sending only 10 or 20 values from the DCT array (the rest of them are zeros). That saves us a lot of transmission bandwidth, or space, if we're storing the video image.

DC coefficient (the "number" around which all the other coefficients vary)

end of block

If the coefficients are read out in this way, most of the final numbers will be ZEROS, and they can be sent by a unique code word, taking up even less space.

Note that this is a block of "frequencies" NOT pixel "values".

How a block of DCT coefficients are read

To further complicate this understanding of how DCT works, the value array is read in a zigzag fashion, so that the likelihood of obtaining long strings of zeros is greater. A long string of zeros, in turn, is sent by a unique code word, which takes less time than sending a series of zeros one after another.

One of the neat things about the DCT process is that the compression factor can be determined by the values contained in what's called a "normalization array," which is used in the actual DCT calculation. Change the array, and you can change the compression factor.

The DCT transform, as just described, is used in both JPEG and MPEG compression systems.

What Makes MPEG Different?

JPEG is for stills; MPEG is for motion. With DCT, we've dealt with video on the single-frame level, but MPEG goes a step further. MPEG uses prediction. MPEG encoders store a sequence of video frames together in digital memory, and have the ability to look "forward" and "backward" over this short time period, to see changes from one frame to the next.

Forward prediction uses information from previous frames to create a "best guess" for the spatial information within the frame we're working on. Backward prediction does the opposite—it uses information from frames "yet to come" and creates a "best guess" for the present frame. Bidirectional prediction is a combination of the two.

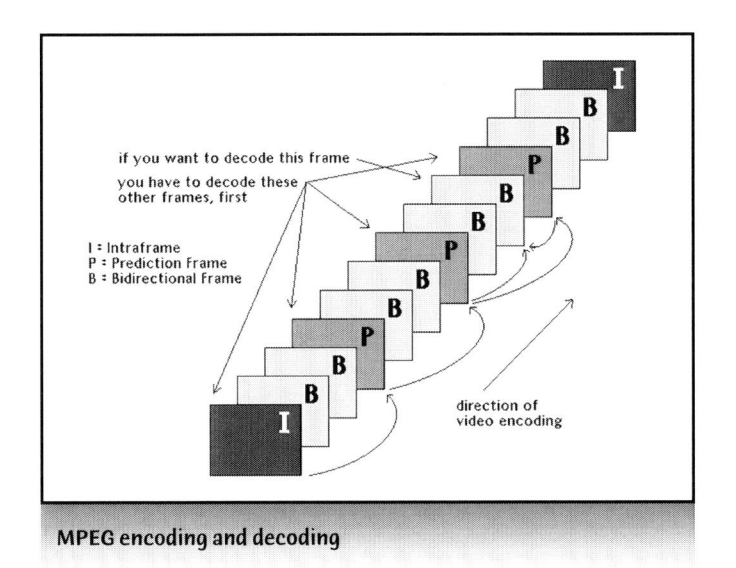

MPEG's frame stream has three kinds of frames within it: Intraframes (which are like a JPEG still); Prediction Frames (those predicted from Intraframes); and Bidirectional Frames (those predicted from both Intraframes and Prediction Frames). These frames are usually referred to by their abbreviations: I, P, and B frames.

I-frames don't produce very much compression after the DCT's been done on them. They are considered "anchor frames" for the MPEG system.

P-frames provide a bit more compression because they're predicted from the I-frames.

B-frames provide a lot of compression, as they're predicted from the other two types of frames; they're totally derived from mathematics. If the math is accurate, they'll look just like the equivalent frames in full-motion analog video, once they've been decoded. If the video has been too aggressively processed, or if there is distortion during the transmission of the digital video signal, artifacts will appear.

The whole series of I, P, and B frames together is called a group of pictures (GOP).

MPEG: Not For Everyone?

So, if MPEG is so great, why not use it for in-house production and for distribution everywhere, in addition to transmission?

The answer lies in the way MPEG actually works. With all those I, P, and B frames sequencing through the stream, it's clear that you can't just cut over on a video switcher to a new video source at any point. Nor can you edit MPEG easily, for the same reason. We only get a new group of pictures approximately every 1/2 second, which isn't often enough for making an edit on the nearest video frame (1/30th of second.)

The first solution designers came up with was to decode the MPEG bitstream to digital or even analog video, do the editing or cutting, and then recompress the creative output back again. This resulted in multiple compressions, with additional layered compression artifacts in the video with each generation. Today, splicing algorithms allow us to make cuts within two MPEG streams by comparing the two streams, finding I-frames in each of them, and aligning the cut points at those moments of coincidence. These systems can also speed up or slow down the bit streams, or even convert the streams so they contain more or different GOPs (with their associated I-frames), making the match points line up for a cut, when we need to do so.

What is a Video Server?

A server, in simple terms, is a system by which we can store large amounts of information and get at it fairly readily. An example of a technologically light analog server would be a large room with many shelves to hold the videotapes, a card catalog or computer database, a tape librarian, and a ladder to reach the shelves. This system gives us mass storage, cross-referencing, and relatively immediate retrieval.

However, in our quest for faster and easier access to material, we developed computer controlled videotape server systems. The earliest broadcast television video servers were analog VTR commercial spot players, first developed in the 1970s (e.g., Ampex ACR, RCA TCR). Later versions incorporated more modern tape formats, more VTRs, a larger tape storage capability, and a robotic system that physically loaded tapes into the videotape machines (e.g., Sony Betacart, Flexicart, and LMS systems.) Eventually, automated tape systems incorporated digital videotape as their format.

Today, hard drives are capable of storing many terabytes of information, and with or without compression techniques, the viability of using hard drives as storage devices for video is no longer in question. MPEG and other compression types allow the bit rate to be lowered to a speed that the hard drives can accept. Another frequently used solution is to "spread" the information to be recorded across several hard drives simultaneously, each drive taking a portion of the total bandwidth of the digital video signal.

One of the biggest concerns about computer hard drives is shared by all individuals who have suffered a personal computer's hard drive crash: what about reliability? The most important requirement of a video server is that it always works, 24 hours a day, without a chance of failure, otherwise a station could go off the air. Fortunately, additional stability can be achieved by configuring groups of disk drives into various redundant arrays of independent disks (RAID). The RAID technology also adds error correction algorithms to the system, and with that comes the ability to swap out a defective drive from the array, without losing your video information. In this case, a fresh, blank hard drive is inserted in place of the faulty one, and the information can be reconstructed over a few hours by the error correction and RAID software.

A typical video server

Applications

Video servers are now used in many places in the television system. Commercials originating in the master control of a television station are now automatically played to air without the use of videotape, while that station's newsroom uses a central server system for editing and producing the daily newscast. Tape delay for feeding programming to different time zones within a specialty cable station is now automated, while the cable companies themselves can now offer pay-per-view economically and reliably.

Commercial Playback

Up until fairly recently, the main use of video server technology (of whatever type) was for commercial playback. This solved one of the largest problems with integrating commercial spots for air, which was the fact that there were several video sources (for example, VTRs) required over a few minute period (during the commercial break). Before video servers, the solution to this was a large bank of videotape machines, loaded up separately by an operator, and rolled in sequence to make up the commercial break. Once the break was run, the entire bank of machines would have their tapes rewound, the tapes would then be stored, and a new series of spots would be laced up in time for the next break.

Machines were then invented that had one or two tape transports and a series of cartridges, each holding one commercial spot. In the beginning, these were based on two-inch quadruplex videotape technology.

The alternative to this was to make a "day tape," a reel with all of the commercial spots pre-edited for the day's broadcast. Master control then only had to run one videotape machine for each break. The day tape had the advantage of less machinery being tied up during the broadcast day, but late changes in commercial content couldn't easily be accommodated. To be able to make fast changes in the commercial line-up, a video server system was the natural solution.

A related application to commercial playback is the simple replacement of videotape machines with video server systems for playback of long-format material (e.g., full programs). This is now done in large network establishments where repeating material is aired several times during a day, or over more than one time zone.

News Editing

With increased demand (especially in news services) for several people to have access to the same clip of video simultaneously, it is now common to see news editing systems incorporating a nonlinear, server-based computer system. The process here involves the cameraperson shooting the raw footage and then having the final takes transferred into the nonlinear editing server. This footage, in its digital form, can be viewed by the assignment editor, the reporter, the writer, the show producer, and, of course, the on-line editor, all at once if necessary. These people can call the material up on different workstations located around the newsroom. Depending on the workstation (or editing station) characteristics, the clips can be seen in different resolutions, frame rates and sizes, and qualities. For example, a low-resolution version may be screened for the reporter to do voiceovers, or for the show producer to monitor the edit's progress, while a high-resolution version will be used to actually send the edited story to air.

Time Zone Delay

With the large longitudinal spread of Canada and the United States, a single satellite broadcast can appear in up to five time zones simultaneously. With material that may not be suitable for all audiences at once (for example, "mature content"), separate "east-west" split feeds are quite common. Usually, a constant time delay (typically three hours) is all that's required. A video server (in its simplest form) is ideal for such a situation. The video information from the east feed is sent into the server, stored on the hard drives for three hours, and then automatically played out again for the west feed. The delay is continuous, with no worry about videotapes having to be rewound or breaking during playback.

Pay-Per-View

Cable TV companies have been offering pay-per-view and video on demand for years, which requires that the outlet have multiple program outputs of the same movie or special event, usually with different starting times. Because hard drive video servers can have several outputs, even reading the same material seconds after each other, it's possible to load one copy of a movie into the system, and play it out simultaneously to many channels.

Almost all of these applications have incorporated into them a comprehensive file management system so that the various parties who have to work with the video files can do so effortlessly and quickly . . . and without a ladder.

7

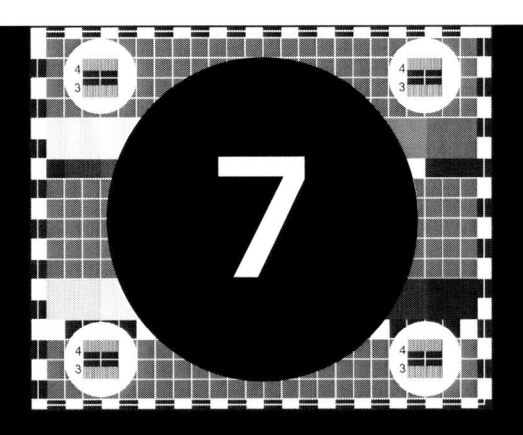

MEASURING VIDEO

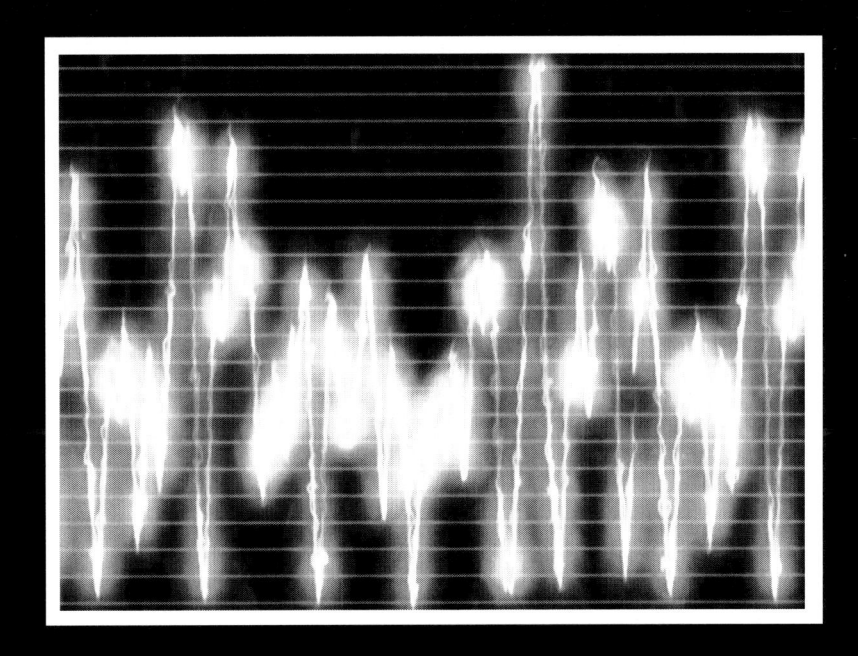

"Don't you wish there were a knob on the TV to turn up the intelligence? There's one marked 'brightness,' but it doesn't work."

—Gallagher

Given the complexity of analog and digital video, it's not surprising that we need special measuring devices to allow us to carefully check the quality of these signals. In this chapter, we will explore ways of monitoring the various attributes of video.

WAVEFORM MONITOR

We have already described in great detail the nature of analog and digital video. So that we may view the precise wave formations in these signals, we need a "waveform monitor."

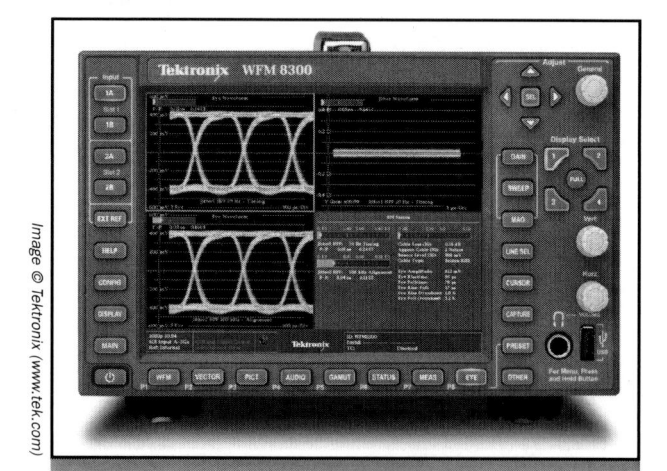

WFM8300 Advanced Analog/SD/HD/3G-SDI Waveform Monitor

Image © Tektronix (www.tek.com)

Old-style waveform monitor (Tektronix RM 529)

For decades, waveform monitors, vectorscopes, and similar devices used cathode ray tubes (CRTs) for their display. While today's signal measuring and monitoring devices commonly use LCD displays, it's important that we understand the theory behind the old methods as well, since the new display devices often mimic legacy technology. So, while in this section there will occasionally be illustrations of CRT displays, keep in mind that the same functions and views are available in newer systems.

In an earlier chapter, we described display devices (such as video monitors) as having screens that are scanned from left to right, and top to bottom. But we don't necessarily have to do it this way. In fact, with the exception of video monitors, CRTs, in particular, are rarely scanned this way.

We can think of a waveform monitor as a sort of "programmable CRT." Its left-to-right scanning pattern is very similar to that of a picture monitor, except that we can change the speed of that scanning. The vertical scanning, however, is a representation of the voltage present at the input terminal of the waveform monitor (i.e., the video signal we're evaluating). Video from a camera or other source normally goes from the bottom of sync at −40 units up to 100 units of peak white level. When looking at a waveform display, the brighter the video level is at a particular moment in time, the higher it appears on the face of the waveform monitor screen. The entire analog video signal carries about 1 volt in electrical energy—a little less than what you would find in a flashlight battery.

Keep in mind that, while switching the waveform monitor's controls in various ways, you are only changing the display of the monitoring device. You aren't in any way affecting the video source itself—just how you're looking at it.

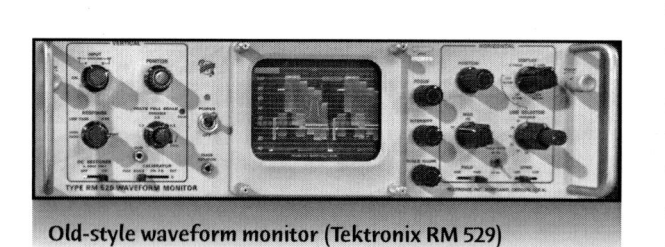

INPUT CONTROLS

Filtering

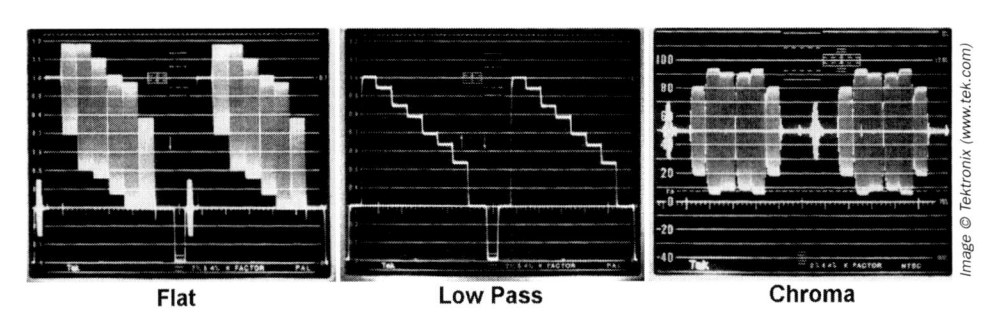

| Flat | Low Pass | Chroma |

This is the switch that allows us to vary the frequency response of the video amplifier in the waveform monitor. There are three positions available to us: FLAT, LOW PASS, and CHROMA. The waveform illustrations are pictures of colour bars with the waveform monitor switched to the various filter positions.

The Flat position allows us to view luminance and chrominance information—the whole range of frequencies in the composite video signal.

The Low Pass position passes just luminance information—all the chrominance information has been stripped from the waveform display. The term "low pass" is a bit of a misnomer, since what's really happening here is closer to notch filtering out a particular frequency—colour subcarrier, 3.58 MHz—leaving only the luminance information on the display screen.

The Chroma position is the opposite of the Low Pass—the luminance information is now missing, leaving just the chrominance. This is a band pass filter—only colour subcarrier is passed to the display, while all of the luminance information (including sync) has been removed.

The waveform monitor has all of these filter variations so that we can observe various aspects of the video signal, without others getting in the way. For example, sometimes we want to see just the luminance of a video signal (e.g., to check complex video levels), but the picture content may have a lot of highly colour-saturated areas in it. This extra chroma information obscures our analysis of the black and white details. In this case, the Low Pass capability is used to eliminate the colour details in the waveform display.

On most new waveform monitors, holding down the filtering switch selects an "alternating" mode. In this position, the Low Pass mode will be turned on for the left side of the monitor display, and the Flat mode will be selected for the right side of the display. The switching between these two modes occurs during the horizontal or vertical sync interval and allows us to view the video signal in two different ways simultaneously.

Reference

Most of the time, the waveform monitor is set to sweep its display, left to right, in synchronization with the video signal being monitored. This means that we can see lines or fields of video clearly, with no movement or drift in the display.

Occasionally though, we want to see any small amounts of movement or timing errors if they are present. This is often used, for example, when "timing" sources to an analog switcher, so that dissolves, wipes, and keys can be generated cleanly. In instances like these, we can have the waveform monitor locked to an external synchronous reference, usually from an "external sync" signal source. This reference usually comes directly from the sync generator at central racks. The sweep of the waveform monitor will now be locked to the external, stable reference, and so any instabilities in the signal being monitored will show up as a movement of the waveform. This can be measured, and steps can then be taken to deal with it as required. We will discuss this separately in the section on video timing.

Input Selection (A/B)

As an operational convenience, waveform monitors come with a switch to select between two different video sources. The switch is usually called an "A/B" switch. On newer monitors, holding down the switch for more than a moment selects a special "alternate" mode where both input signals can be viewed simultaneously on one waveform display. The switching between these two signals occurs during the horizontal or vertical sync interval, similar to the dual Low Pass and Flat filtering mentioned above.

VERTICAL ADJUSTMENTS

Gain

The gain of the video amplifier can be varied, similar to the volume control on an audio amplifier. Obviously, the video signal itself doesn't get "louder" or brighter; instead, it takes up a proportionately greater vertical area of the waveform screen. Usually, there are two controls that you can use to adjust gain. The first is a switch to go between a UNITY (normal, calibrated) state, a five times magnification (X5) mode, and a variable (VAR) mode. In UNITY, the waveform display acts normally—one volt of video will usually land between the −40 and 100 markings on the graticule (the lines on the waveform monitor's screen). In the ×5 magnify mode, the signal will be five times as high as normal. With the VARiable mode, the second control—a GAIN knob—comes into play, to allow you to vary continuously how large the signal will be displayed on the screen.

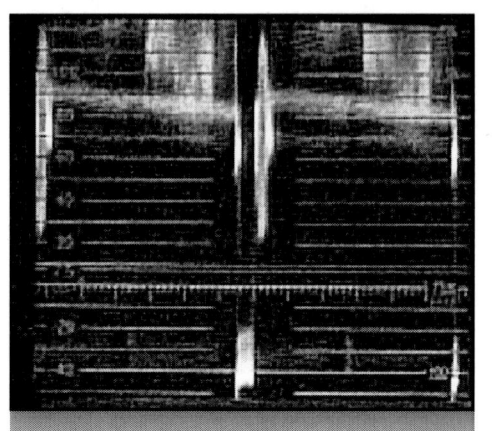

Two line display, with GAIN turned on

If UNITY is a reliable, calibrated, and accurate mode to measure video signals, why would we want to change this? Often, engineers want to measure very small anomalies in video signals—for example, video noise or hum—and these problems are difficult to see in the UNITY setting. If we switch to ×5 mode, for example, these details are five times larger, and are therefore easy to measure on the display and monitor.

DC Restoration

Analog video is a high-frequency alternating current (AC) with a frequency range of 30 Hz to 4.2 MHz. The video input to a waveform monitor is generally "AC coupled," which means that any direct current (DC) voltage offsets will be eliminated the moment the video enters the monitor. This is done so that the video signal doesn't go off the top or bottom of the scale (which would be easy, considering that we're measuring a voltage as small as 1 volt). This AC coupling causes a problem, however.

Since the video is AC coupled, the waveform has a tendency to "float" up and down the monitor screen, in an attempt to centre itself within the middle of the display, as the video level fluctuates. This is not useful when trying to measure video levels, since the entire signal wobbles up and down the display, making it impossible to take accurate readings. To solve this problem, the waveform monitor has a "DC restoration" circuit built into it. This circuit looks for the sync pulses of the video signal and vertically locks the waveform display to it, regardless of what the overall peak-to-peak video level is. Since the signal no longer floats up and down, we can use the vertical position knob to position blanking at 0 units on the display's graticule. Once the waveform monitor is properly set up like this, accurate video levels can be read. Most of the time, therefore, you should leave the DC restoration ON.

Position

This is a variable control that positions the waveform display up and down on the screen. It is normally adjusted so the blanking of the video waveform rests at 0 units on the graticule.

HORIZONTAL ADJUSTMENTS

Sweep

The sweep speed from left-to-right can be varied on a waveform monitor. Normally, we will want to view either one or two lines of video, or occasionally two fields, across the waveform monitor screen. The sweep switch is usually labelled 2 LINE and 2 FIELD. Some waveform monitors also have a selection here for 1 FIELD and 1 LINE.

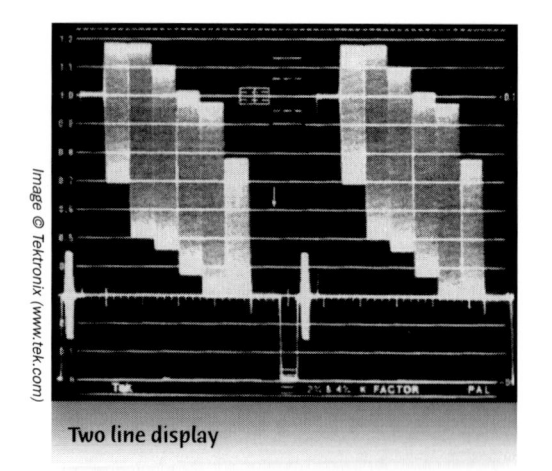

Two line display

The 1 LINE position gives a fast display of video lines 1, 3, 5, and so on, followed by lines 2, 4, 6 and so on. This unusual line number sequence occurs because the NTSC video signal is interlaced. This display is sweeping horizontally at a rate of 15,734 Hz. Therefore, all of the lines of a video signal merge into one on the display. The 2 LINE position gives a similar display, but since there are two lines visible at any one moment, the display is sweeping at a rate of 7867 Hz (15,734/2) and shows you lines 1 & 3, 5 & 7, and so on, then 2 & 4, 6 & 8, etc.

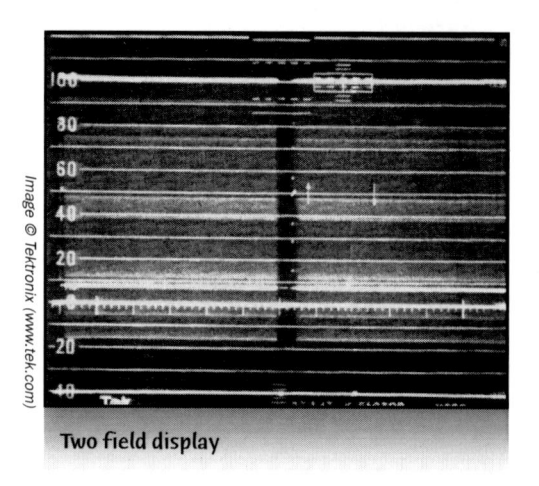

Two field display

The 2 FIELD position shows you the first and second fields of a frame of video, with a vertical interval in between them. The sweep speed is approximately every 30th of a second, which is also why you sometimes perceive that a 2 field waveform monitor display is flickering a little bit. There is no 1 field display setting on a waveform monitor.

Magnification

The horizontal sweeps can also be magnified, similar to the vertical GAIN. Generally, the MAG control is a two-position switch: ON and OFF.

Once turned ON, various details can be displayed. During the 2 LINE display, the magnification "stretches" the display horizontally. You will notice on the waveform graticule, at the 0 units line, a series of larger and smaller tick marks across the screen. Each one of the major marks, in the horizontal magnification mode, represents 1 microsecond (1 millionth of a second). This MAG setting is commonly called "1 microsecond per division (1 μs/div)" and can be used to very accurately measure small areas of the horizontal interval, sync pulse widths, colour burst cycles, and so forth.

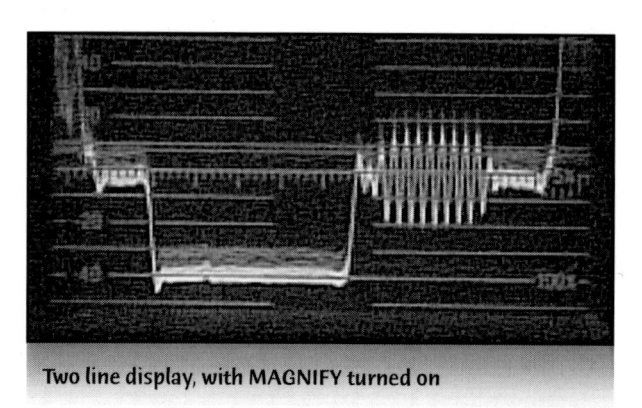

Two line display, with MAGNIFY turned on

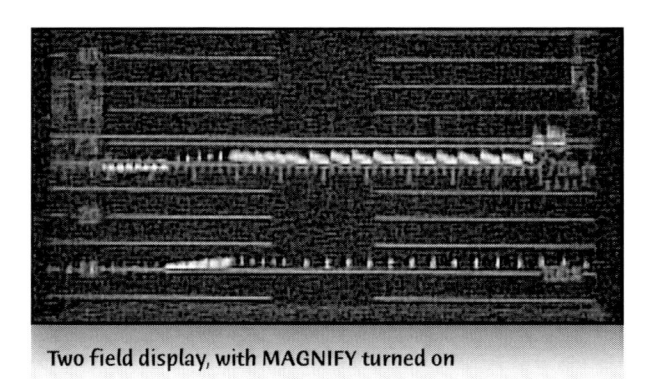

With a 2 FIELD display, the MAG mode shows you one full vertical interval (the one between the two fields mentioned earlier). This is useful for checking that the vertical sync and equalization pulses are being produced accurately, checking for signals inserted in the vertical interval, and making other critical observations in this portion of the video signal.

Two field display, with MAGNIFY turned on

Field Selection

There are two slightly different versions of the vertical interval (they differ for field one and field two, due to the interlace pattern of NTSC video). We need a way to select which field we wish to observe (or which vertical interval, when using the horizontal magnify mode). The switch to select this is the FIELD switch, and is labelled appropriately with a FLD 1 and FLD 2 position.

Position

This is a variable control that positions the waveform horizontally. It is used to move the waveform left or right to centre the display for the portion of the video signal you need to observe for accurate measurements.

DISPLAY ADJUSTMENTS

Focus

If you are using a waveform monitor with a CRT, this knob focuses the CRT beam for optimum definition. It is not found on waveform monitors using an LCD display.

Scale Illumination

This controls the level of graticule illumination (on a CRT display) or the level of discernability (on an LCD display).

Intensity

This controls the intensity beam (on a CRT display), or the level of discernability (on an LCD display). It's similar to the brightness control on a picture monitor.

LINE SELECTION

In addition to all the functions mentioned above, the waveform monitor can display a single particular line of the video signal, in either field. Selecting LINE SELECT enables this feature.

In a 1 LINE or 2 LINE sweep mode, you will be able to see the particular line in the video signal you selected for monitoring. The line and field information is printed on the waveform monitor screen with a built-in character generator. On many units, an external video output at the back of the waveform monitor allows you to hook up a video display that will show a superimposed bright line within the picture, showing which video line you're monitoring.

GRATICULES

The NTSC graticule has two main vertical scales. The left side of the scale is marked in IRE Units and extends from −50 to +120 in 10-unit increments. Black level (also called "setup") is represented by a dashed line at 7.5 units. There are small ±2 and ±4 unit "tick marks" at the bottom centre of the scale, around the −40 units line, for measuring small changes in sync amplitude.

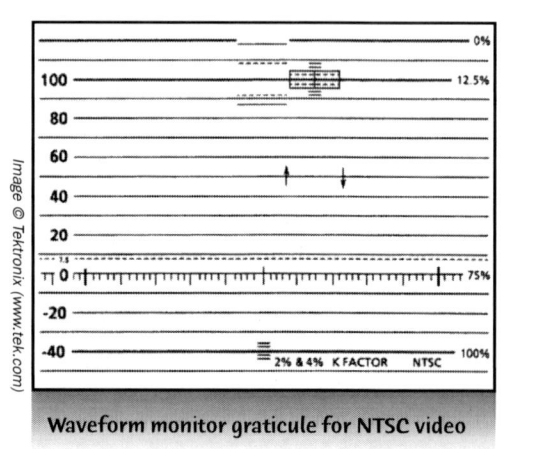

The percentage scale on the right of the graticule (going from 0% at the top of the screen to 100% at the bottom) is for measuring "modulation depth" and is used by engineers to monitor analog NTSC transmission characteristics.

The horizontal reference line is the baseline at 0 IRE. This timing line is 12 divisions long and takes on different timing intervals depending on the sweep rate selected. In 2 LINE, each major division is 10 µs (as we saw before, µs means microsecond, a millionth of a second); when magnified, each becomes 1 µs. In 2 FIELD, the scale is not used, since this is mostly a general view monitoring mode.

Waveform monitor graticule for NTSC video

When using the waveform monitor, the sync tip of the video signal should be just touching the −40 IRE marking. This being so, the 0 IRE baseline corresponds with 0 units of video; 7.5 units of video (black level, or setup) should be on its corresponding line. The white flag of colour bars (if that is the video source you're monitoring) just touches the 100 IRE scale marking.

Vectorscopes

The visual sensation of colour is described in three qualities: luminance, hue, and saturation. Luminance is the brightness of a video signal. This is handled by the luminance (black and white information) portion of the NTSC video system and is easy to measure with a waveform monitor. Chrominance in a video signal is determined by the phase and amplitude of colour subcarrier. Phase represents the hue of a colour. The saturation of a colour is the degree to which a colour is diluted by white light to distinguish between vivid and weak shades of the same hue. For example, vivid red is highly saturated; pink has little saturation. In life, saturation is a product of both luminance and chrominance combined; in analog colour television, colour amplitude is determined by colour subcarrier level only.

White, black, and grey, of course, are not colours and therefore have no phase associated with them, nor do they have any chrominance level.

The hue and colour amplitude information in the NTSC analog colour television system is carried on a single subcarrier frequency: 3.579545 MHz. This signal, in modulated form, is called chrominance. The hue information is carried by the subcarrier's phase, and is measurable as being anywhere between 0 and 359 degrees. This is reminiscent of the circular protractor you might find in a high school math toolkit. The colour amplitude information is carried by the level of the subcarrier at any given moment in the composite video signal. Finally, a reference signal—a short colour burst, of several cycles of 180-degree subcarrier at the start of every line—is sent within the video signal. Clearly, the most logical way of looking at a particular colour's NTSC hue (phase) and amplitude (saturation) is with a circular display.

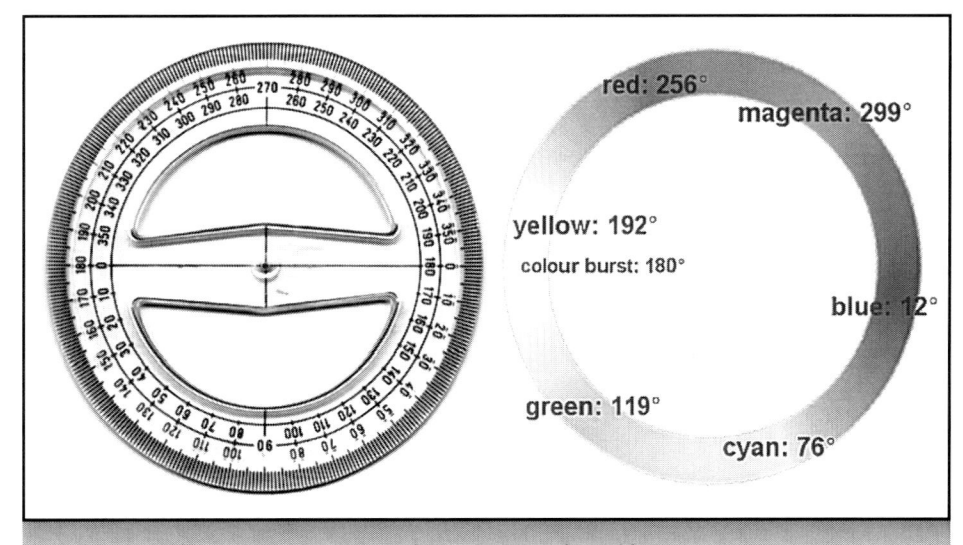

Protractor angles compared to various hues in NTSC colour video

Image © Tektronix (www.tek.com)

Vectorscope display, showing NTSC colour bars

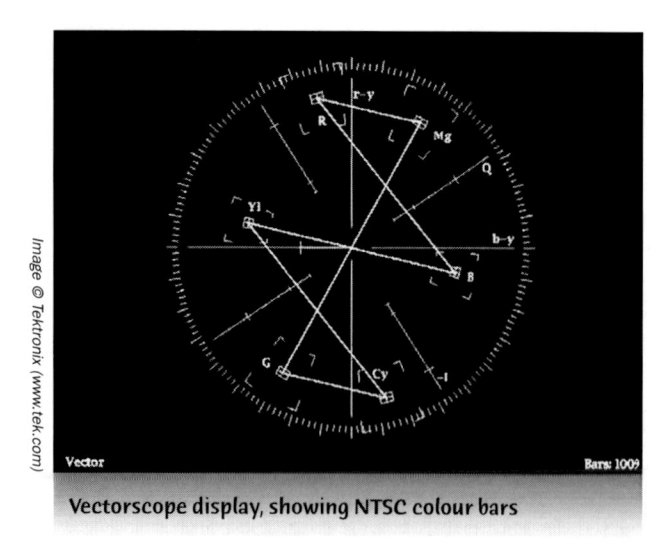

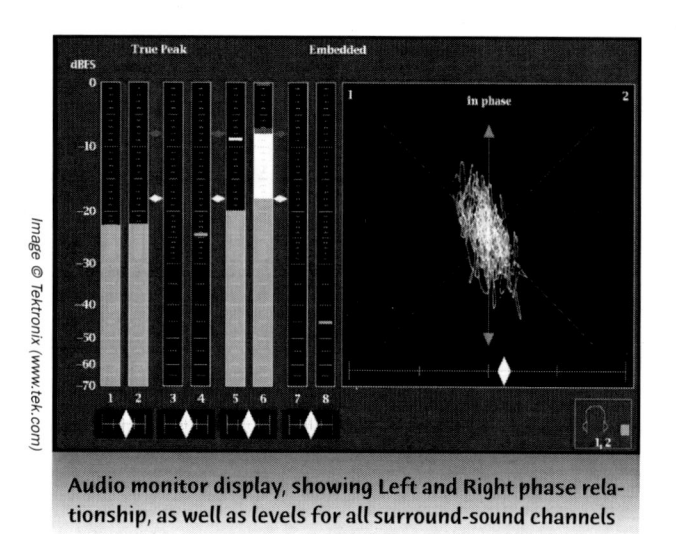

Image © Tektronix (www.tek.com)

Audio monitor display, showing Left and Right phase relationship, as well as levels for all surround-sound channels

A vectorscope takes a composite video signal, samples the colour burst from it, and uses that as a reference for the display. The remainder of the colour information, on a line by line basis, is decoded in a way similar to a conventional colour monitor. Two new signals are created: one called R-Y (the red channel, mixed with an inverted luminance signal) and the other called B-Y (blue, mixed with inverted luminance). The R-Y is sent to the vectorscope display and shows its variations in a vertical orientation; the B-Y is also sent to the display and its fluctuations are shown horizontally. The two variations in the two signals combine together to produce vectors—small lines indicating the various colours in a video signal—starting from the centre and extending outwards towards the circumference of the display.

The vectorscope is essentially a simple display with the capability of exhibiting two signals simultaneously: one on the left-to-right (also known as "X" axis) and the other on the top-to-bottom (also known as "Y" axis) fluctuations in its display.

Because of this, other two-dimensional signals, such as audio, can be monitored with such a display. Some vectorscopes, therefore, have two audio connections at the back of the unit, so the left and right channels of a stereo signal can replace the normal X and Y chrominance information. This can be used by audio engineers to check the amplitude and relative channel phase of stereo program audio. In modern multipurpose monitoring devices, a dedicated audio display appears showing not only audio phase, but also the relative level of all channels of 5.1 surround sound.

Input Controls

Reference

Most of the time, the vectorscope samples the colour burst from an analog video source and uses that as its display reference. Occasionally, though, we want to see any small amounts of subcarrier phase shift if they are present. This is used when "timing" sources to an analog switcher. In instances like these, the vectorscope can be switched from INT to EXT, and locked to an external continuous subcarrier source, usually from a sync generator at central racks.

A/B

As an operational convenience, vectorscopes, like waveform monitors, come with a switch to select between two different video sources. On some devices, you can view both signals simultaneously; however, this usually requires an external source of stable subcarrier.

Gain

Variable

This is a switch and knob combination that allows you to vary the length of the vectors on the display. This is useful for accurate setting of colour bars and external subcarrier measurements for timing purposes. Normally though, the GAIN is left in its accurately calibrated UNITY position—switched off.

Phase

This is a continuously rotating knob with a 360 degree range to set the phase of the decoder reference. This means that you can rotate the vectors around the screen to any degree setting that you want. It is usually set so that the colour burst on the display is pointed to 180 degrees—directly to the left, or 9 o'clock, position.

Display Adjustments

The display adjustments on vectorscopes are similar to those available on waveform monitors.

Focus

If you are using a vectorscope with a CRT, this knob focuses the CRT beam for optimum definition. It is not found on vectorscopes using an LCD display.

Scale Illumination

This controls the level of graticule illumination (on a CRT display) or the level of discernability (on an LCD display).

Intensity

This controls the intensity beam (on a CRT display), or the level of discernability (on an LCD display). It's similar to the brightness control on a picture monitor.

Vectorscope graticule for NTSC colour video

Detail of vectorscope colour bar 'target'

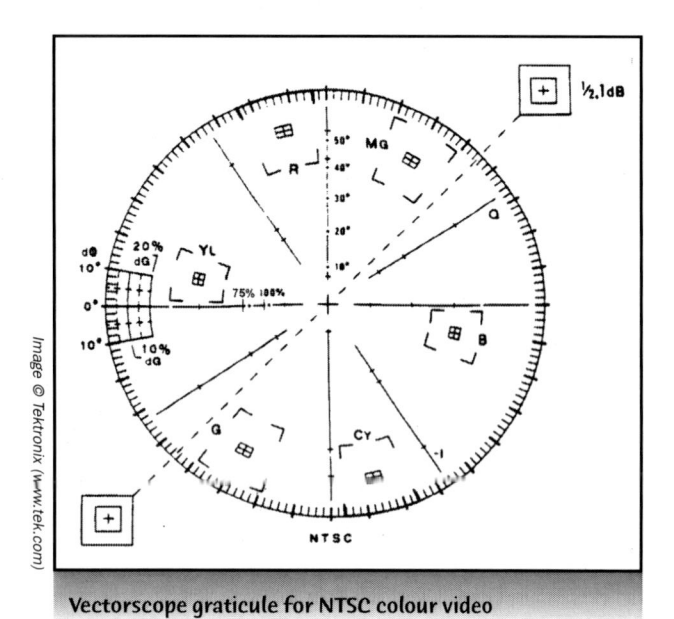

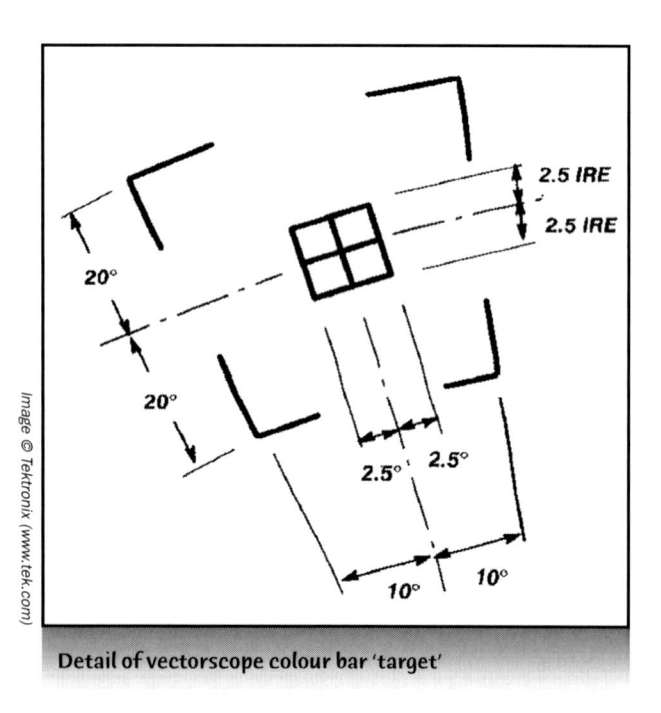

Graticules

On the vectorscope scale, the polar display indicates the degree of phase (hue) by the relative position of the vectors around the display with respect to the colour burst. The relative amplitude of chrominance (saturation) is shown as the displacement from the centre towards the outside of the display.

There are a number of small boxes within the vectorscope display. Each shows where a particular bar from a colour bar test pattern should fall, when equipment is lined up correctly. There are actually two boxes for each colour—a smaller one inside a larger one. The large box represents ±10 degrees of the correct phase, and ±20% of the right chrominance level. The smaller boxes represent ±2.5 degrees and ±2.5 IRE units for each colour. The idea, of course, is to get the vectors of each colour bar as closely as possible into the smaller boxes. This is done by adjusting the phase and chroma level controls of the device being monitored (video playback unit, camera encoder, etc.).

When using the vectorscope to measure the phase of two channels of audio, various Lissajous patterns are shown. If there is no phase error between the channels, the display will be a diagonal line, at a 45-degree angle running from the bottom left to the upper right of the display, and the peak audio levels will fall into those two boxes on the display shown above. When the signals are partially out of phase, the pattern will be displayed as an ellipse. When the audio signals are 90 degrees out of phase, the pattern is a circle. At 180 degrees out of phase, the display is again a straight line, but the axis of the pattern is rotated by 90 degrees—it runs from the upper left to the bottom right.

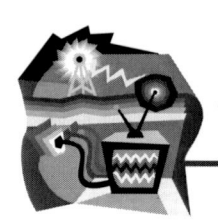

ANALOG VIDEO TIMING WITH WAVEFORM MONITORS AND VECTORSCOPES

As each video source enters a production switcher, its video scan must begin and end in time at precisely the same moment as any other source. If this is not so, you will get a horizontal shift in the picture as you try to dissolve or wipe from one source to another. This is because the switcher holds the horizontal sync of the first source until the transition is completed—the jump happens as the switcher suddenly references to the second video source's horizontal sync signal. If the scans begin at vastly different times, some switchers will simply cut from one source to another—they won't allow the dissolve or wipe to take place at all. This alignment of the scanning rates is called "horizontal" or "H" timing, and it affects the horizontal position of one video image relative to the other's.

Because we are working with colour television signals, each video source's colour burst must be in phase with every other source's burst. If this is not the case, you will get a colour shift when dissolving, e.g., a scene will go from normal looking colour to "purple" or "green." This is called "system phase" or "subcarrier phase." This is not to be confused with what's known as "burst phase," which is the local colour "hue" setup at the video playback machine's processing amplifier or time base corrector.

Frame Synchronizers: A Side Note

Many video sources are "free running." That is, their scans bear no relationship in time to our production facility's sync generator. A good example of this is a satellite video feed. The satellite is thousands of kilometres in space, and it's sending back to earth a video signal running on its own timebase. It's professional quality video, certainly, but the horizontal and vertical synchronization pulses (and the colour burst) can begin anywhere relative to other equipment within a television station on earth. Such a source taken directly to air would be objectionably shifted horizontally (and possibly vertically as well), and the colours would constantly be shifting through various hues.

Our solution to this problem is the frame synchronizer. This box takes an entire frame of the remote source's video (whenever it may have started its own scan), and stores it in digital memory. It then releases it when cued by the in-house sync generator's next vertical interval. This makes satellite, microwave, laser link, and domestic camcorder sources compatible for broadcast use.

This also seems like a perfect solution to all of our timing problems. Why not use frame synchronizers for every source in the building, at every input to every switcher? First, it's expensive to purchase that many frame synchronizers, if we have an easier way to set up proper timing. As well, every time you feed a video signal through such a device, it is converted to digital form and then back to analog again. This degrades the signal somewhat. Also, this process results in the video being delayed by up to one frame (about 1/30 second) with respect to the audio, which has not been through any processing. While a one frame delay may not be that noticeable to us, doing this two or three times (e.g., incoming to the production facility, then to a production switcher, then to the master control switcher, and so on) results in unacceptable video to audio time shifts.

How to Time a Video Source

The TD (Technical Director) has access to a waveform monitor and a vectorscope. Both devices are "externally synchronized" for this operation—they have a constant external reference to which they can refer. For timing video sources in an

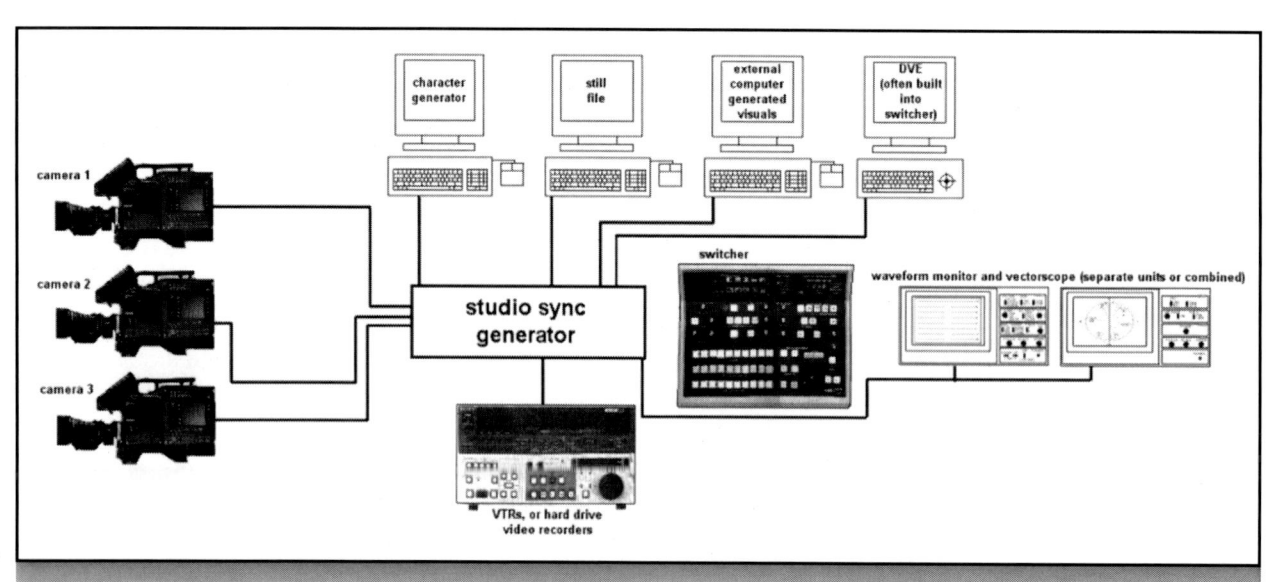

Typical studio configuration, showing distribution of sync pulses to various pieces of equipment

analog switcher, the TD must ensure that the waveform monitor's EXT/INT switch is set to EXT (with a suitable external sync source, from the station's sync generator). The vectorscope's EXT CW (external continuous wave subcarrier) selection is also selected (with a suitable source of continuous wave subcarrier, again from the sync generator). All video sources entering the switcher must also be referenced to external sync.

The waveform monitor and the vectorscope should be set up to monitor either the switcher's program output bus (the row of switcher buttons where each video source is selected to go to air), with any processing amplifier bypassed (if the switcher is not live on air, of course), or the switcher's preview bus (the switcher buttons where the next source to be taken to air will be previewed). These are usually prewired by the television station engineers to come up as the "A" side and "B" side of the waveform/vectorscope video inputs and selection switches.

Once this condition is met, the waveform is set to 1 μs/div, so an expanded horizontal sync pulse is in view. The vectorscope's burst should be set to 180 degrees (straight to the left). Both of these settings are prepared while "BLACK" is selected by the relevant switcher bus, since this source is internally generated and locked by the switcher equipment. If desired for easier viewing, the vectorscope's gain can be increased.

To check horizontal timing at a video playback device, for example, the TD will switch between BLACK and the source in question, back and forth. The video playback operator is "talked through" adjustments of the horizontal timing pot on the video playback machine, so the TD sees no horizontal shift between the two, while viewing the expanded horizontal interval on the waveform monitor. As the actual length of the horizontal interval width may vary slightly from source to source, the TD uses the right edge of sync as a reference—this is where each video line's blanking starts.

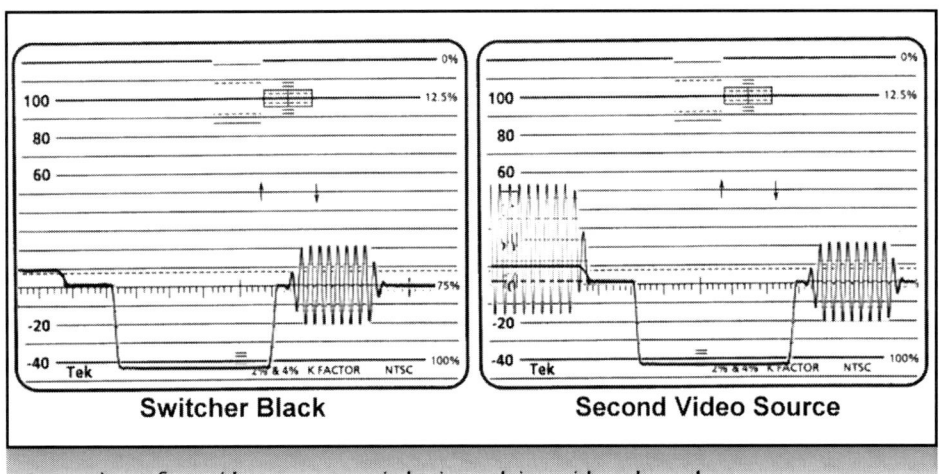

Comparison of two video sources not in horizontal time with each another

To check subcarrier timing, the TD will select the source in question, after having set the vectorscope's burst phase to 180 degrees on BLACK. He will "talk through" to the playback operator who is adjusting the colour subcarrier pot on the machine, until the new source's subcarrier phase is also at 180 degrees.

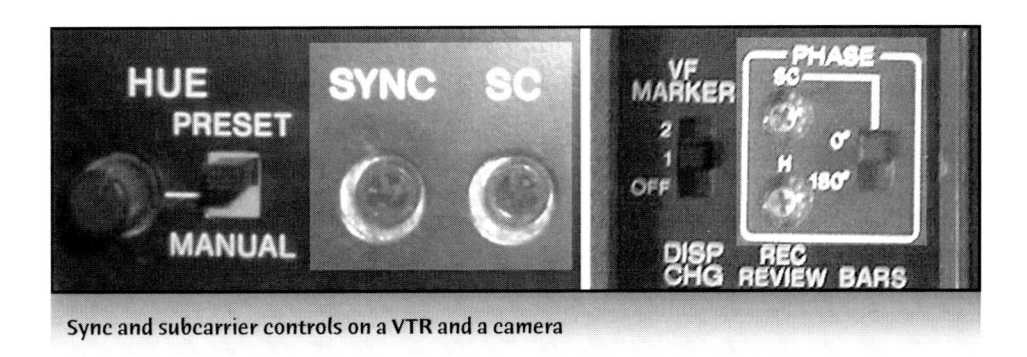

Sync and subcarrier controls on a VTR and a camera

The video playback device is now timed to the switcher and anything played back from that machine can be dissolved or wiped. It also will have the correct colour phase. Important: Once all timings are completed, the TD must remember to turn the processing amplifier back on, if it was bypassed for the timing operation.

Cameras and other timing-dependent sources can be set up similarly. This entire operation should take only a couple of minutes at most.

ANALOG VIDEO VERTICAL INTERVAL INSERTIONS

The vertical interval features several important elements to ensure that the horizontal synchronization does not go out of alignment with regard to the interlace pattern of NTSC television. Specifically, there are two sets of equalizing pulses

Typical off-air vertical interval, showing insertions

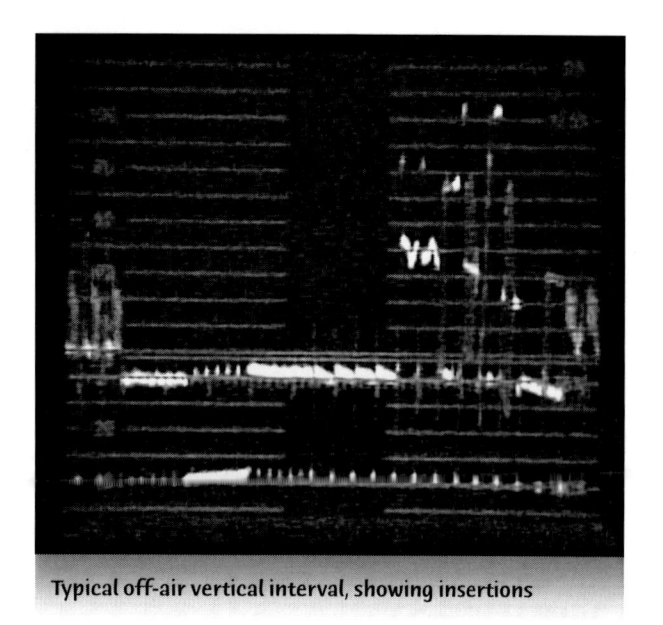

generated during the vertical interval, along with the vertical sync pulse itself, of course.

But, these special pulses only take up 9 horizontal lines' worth of time; active video doesn't start until much later, on line 22. Between the sync and equalizing pulses, and the first line of active video, there are several lines that consist of only blanking, a horizontal sync pulse, and colour burst. Originally these blank lines were inserted into the video signal to minimize the visual degradation perceived on early television sets, as the receivers' electronics settled down from the vertical retrace process. Today's television sets and monitors are much more stable in this regard, so the blank lines can now be replaced by other signals, used for various special purposes.

These signals are generally referred to as VITS (Vertical Interval Test Signals.) This is a general term incorporating test signals inserted within lines 10 through 21. These signals vary according to the needs of the television station, and the country of transmission.

VIRS stands for Vertical Interval Reference Signal. It consists of a 70-unit IRE signal modulated with subcarrier of the same level and phase as colour burst, with 50-unit and 7.5-unit IRE level signals. It is normally used for either manual or automatic adjustment of chrominance gain and phase, luminance, and setup parameters of a video signal. The modulated 70-unit signal is of average chrominance phase, at average Caucasian skin-tone luminance level. The 50-unit IRE signal represents an average picture luminance level, and the 7.5-unit IRE pedestal is used for picture black level setup. VIRS is normally found on line 19 in both fields. Some analog television sets can make use of this signal to align themselves automatically for the best picture quality. For this to work, each source is supposed to generate its own VIRS signal, and shouldn't be stripped out or regenerated later—the idea being that, as you go from source to source, the viewing or monitoring system adjusts for each one. This concept has never really caught on, however.

GCR—Ghost Cancelling Reference—is a signal transmitted on line 19 of some television stations. It's a sweep in frequency from 0 Hz to 4.2 MHz, occurring over one video line. It's only useable by more recent analog television receivers that have ghost cancelling ability. The GCR is transmitted with the regular TV picture and is compared to a clean version that resides in the television set. Any differences found are used to tune out ghosts in the received transmission.

VITC is Vertical Interval Time Code. As well as being recorded on videotape as audio in a linear fashion, time code also can be converted to video information. Many time code readers and generators are capable of reading and writing this data in the vertical interval on a given line (line 19 is often used).

Closed Captioning is a method by which text information can be received by individuals who have a special decoding box or any television set measuring 14" or more in screen size. This is used to transmit text captions on-screen for the hearing impaired. The captioning data signal is contained in the television signal vertical interval on line 21, field 1. The signal must be transmitted from a station intact and on the right line and field, or consumer decoders will not operate. The captioning data signal is a legal requirement in many television signals, and is as much a part of the program as the video and audio portions.

Other general test signals such as colour bars, multiburst, NTC7 Composite, FCC Composite, Sin x/x, or frequency sweeps of various patterns can also be introduced into different lines in the vertical interval.

Colour Space, Colour Gamut, Illegal and Invalid Colours

Cameras and other traditionally analog devices usually have video limiters built into them to prevent the creation of excessive luminance and chrominance levels. With the increasing use of computer-generated video, as well as the now common conversion between analog and digital, and between various digital formats, it's now possible to accidentally create illegal or invalid colours within a particular gamut.

Definitions

- *Gamut* is defined as the range of voltages used to display colour. It can also refer to the range of colours themselves that can be displayed within a given electronic visual system (e.g., NTSC television, digital television, computer systems, etc.)

- An *illegal* colour is one that strays outside of a particular video device's allowed gamut.

- An *invalid* colour is one that, while it can be validly displayed in one system (for example, a computer graphics display), will not translate properly to another colour system (for example, a video recording device, or a video's television transmission).

Here's an example of such a problem. If you generate a deep magenta colour in a computer graphics program that is eventually going to be broadcast, there may be many conversions before that happens. If you generated the graphic in SD, it could be upconverted to HD, or vice-versa. It may even be eventually converted to analog video. Your beautiful colour looked perfect on your computer monitor (it would have been a legal colour in your computer's display gamut). Unfortunately, it was an invalid colour, and therefore it got compressed during a subsequent conversion to another format, creating a rather muddy magenta when it went to air.

The Diamond Display

The diamond display allows us to see invalid colours before they are sent anywhere else. It is a special combination display of R, G, and B information.

90 Television Technical Theory

Diamond display, showing colour bars

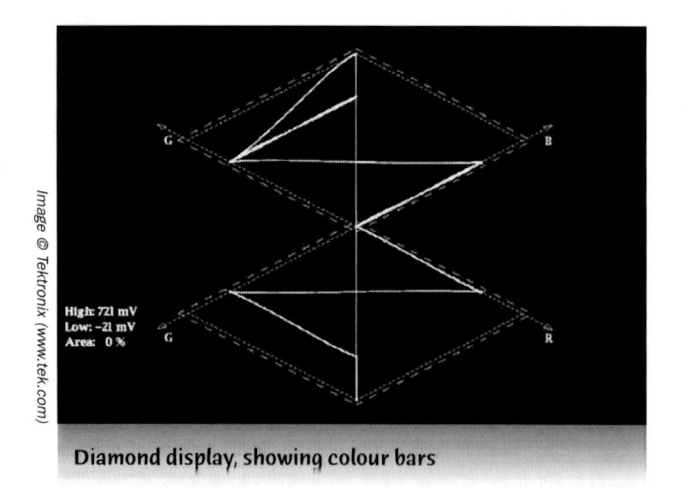

The upper diamond is formed by applying B+G to the vertical axis and B–G to the horizontal axis. This part of the display will indicate all colours in the picture that will be created with a combination of blue and green. The lower diamond is formed by applying –R+G' to the vertical axis and R–G to the horizontal axis, and will show all colours created with red and green. The two diamonds are displayed one above the other, creating the double diamond display.

It's very simple to read, though: for a signal to be in gamut, all of the vectors in the display must stay within the diamonds. If a vector extends outside, it's out of gamut. In the diamond display, black and white signals are simply vertical lines.

Measuring and Monitoring the Digital Television Data Stream

In an analog video system, the signal is a changing voltage that direct represents the video. An analog video waveform monitor and vectorscope make it easy to view the voltage level of the video signal in relation to distinct timing patterns, and to measure the amount of colour information present at any moment in the picture content.

In a digital video system, the signal is a stream of data representing video information. This data is a series of analog voltage changes that must be correctly identified as high or low to be decoded back into video content. The digital signal starts out at a level of 800 mV (millivolts) and can be viewed with a high-frequency oscilloscope or with a video waveform monitor.

In the eye-pattern mode, the waveform monitor operates as an analog oscilloscope with the display swept at a video rate. The rapidly changing data in the transport layer is a series of ones and zeros overlaid to create an eye pattern. With long cable runs, the data tends to disappear into noise—the so-called digital cliff that we discussed in chapter 6. Additional transport layer information such as jitter, rise time, eye opening (extinction ratio), reflections, and data analysis on the received data itself can also be viewed on some digital waveform monitor devices now available.

Eye display, showing good quality digital video stream

Since the data transport stream contains components that change between high and low at rates of from 270 Mb/s (SD digital video) to 2.970 Gb/s (HD digital video), the ones and zeros will be overlaid for display on a video waveform monitor. This is an advantage since we can now see the cumulative data over time, to determine any errors or distortions that might intrude on the eye opening and make recovery of the data difficult.

The digital video waveform display that looks like a traditional analog waveform (baseband video) is really an analog waveform recreated by the digital data, decoded into a high quality analog component video.

Eye display, showing poor quality digital video stream

Eye Pattern Testing

The basic parameters measured with the eye-pattern display are signal amplitude, rise time, and overshoot. Jitter can also be measured. A unit interval (UI) is defined as the time between two adjacent signal transitions. The unit interval is 3.7 ns (nanoseconds) for an SD digital component signal, and 673.4 ps (picoseconds) for digital high definition.

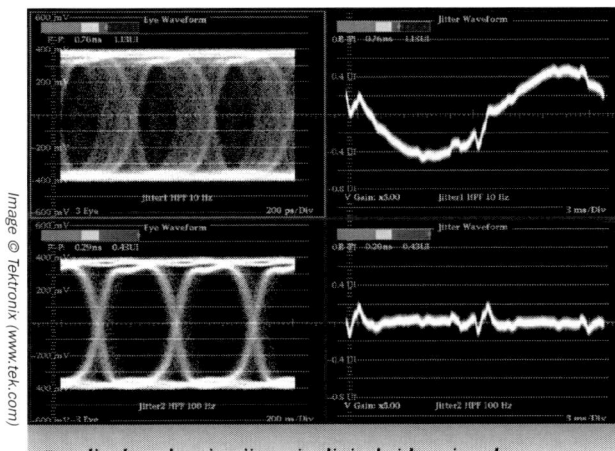

As noise and jitter in the signal increase through a length of video cable, any effect which closes the eye may make the signal harder to decode. Allowed jitter is specified as 0.2 UI. This is 740 ps for SD and 134.7 ps for digital HD. Digital systems will work beyond this jitter specification, but will fail at some point. It's important to maintain these specifications to keep the system healthy and prevent a failure which would cause the system to fall off the edge of the digital cliff.

Signal amplitude is important because of its relation to noise and because of digital video equipment's ability to decode the signal reliably. To read this amplitude, rise-time measurements are made from the 20% to 80% points on the signal. Incorrect rise time could cause signal distortions such as ringing and overshoot, or if too slow, could reduce the time available for sampling within the eye. Overshoot must also be monitored.

Eye display, showing jitter in digital video signal

Most of this digital signal monitoring will be done by the production facility's engineering team. If you are interested in more information about digital signal monitoring, a thorough Internet search will turn up a lot of helpful educational information that will describe these signals in much more detail than can be provided within the scope of this textbook.

8

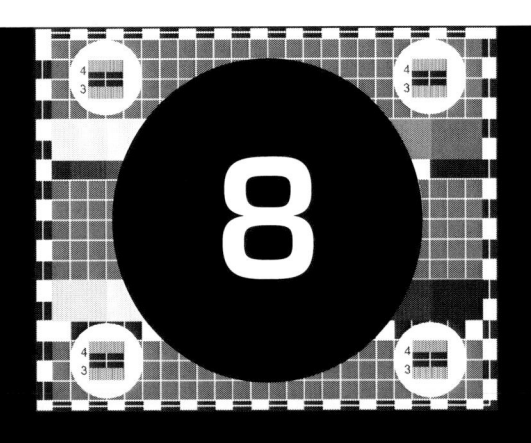

MONITORS AND TELEVISION SETS

"I have a friend who just got back from the Soviet Union . . . the first question he was asked was if we had exploding television sets. You see, they have a problem . . . many are exploding. They assumed we must be having problems with them, too."

—Victor Belenko, MiG-25 pilot who defected in 1976

Picture Tubes

A Legacy Monitor Technology

Historically, television sets and monitors have used a picture tube, ranging in size from one to forty inches across the face of the screen, measured diagonally. The formal name for a picture tube is cathode ray tube (CRT). It gets its name from the essential element of its operation, the cathode ray, which is more commonly called an electron beam. At its narrow end, a typical picture tube used in black-and-white monitors contained an electron gun. Within the gun was a control electrode, connected to the video signal, which accelerated and slowed down the electron beam to recreate the brighter and darker portions of the image. For the electron beam to operate like this, the picture tube was under a very high vacuum, and therefore had to be capable of withstanding the pressure of the outside air. This meant that the glass in a picture tube was fairly thick and, as a result, picture tube monitors were generally very heavy.

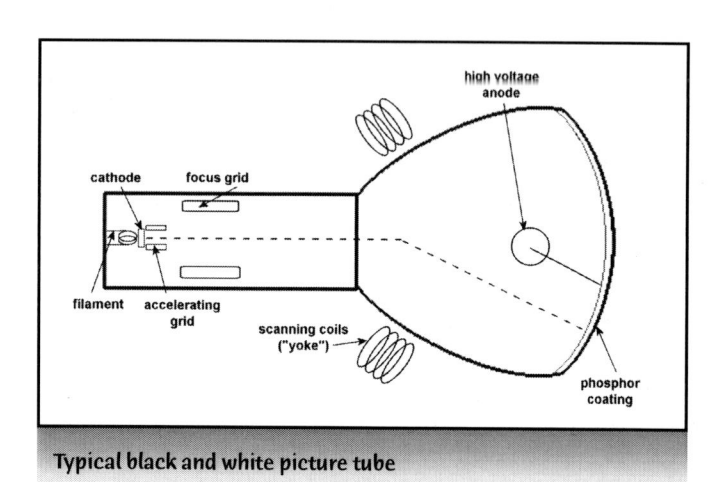

Typical black and white picture tube

The focusing coil or grid, within the electron gun, kept the electron beam narrow, so when it struck the faceplate at the wide end of the tube, it was still very small. The scanning coils were located where the tube just began to expand in size. Rapidly moving sawtooth wave voltages, produced by the monitor's electronics, were passed through these coils, producing magnetic fields that deflected the electron beam in a rectangular, interlaced pattern.

At the wide end of the tube was the viewing screen. It consisted of two coatings. The first, deposited on the inner surface of the glass face, was a material known as a phosphor. The second was a thin coating of aluminum on top of the phosphor. The phosphor was a complex compound, typically containing oxides of sulphur and zinc, which glowed with a bluish white light when struck by high-energy electrons from the beam.

To assure that the electrons in the beam had enough energy, a high voltage (typically 10,000 to 20,000 volts) was applied to the picture tube between the electron gun and the aluminum coating. This positive high voltage at the screen end of the tube attracted the negative electrons emitted from the electron gun. When these electrons hit the screen, they produced a spot of light of variable brilliance, depending on how many electrons were hitting it at a given moment in time.

Black and white cathode ray tube electron gun

This technology, while refined to a high degree of reliability and resolution over several decades, has now been replaced by newer display devices. There are now very few black-and-white picture tube monitors and television sets being manufactured.

Colour Picture Tubes

Colour picture tube monitors and television sets, on the other hand, are still quite prevalent, if only because these monitors will be in use for some time, until they finally expire.

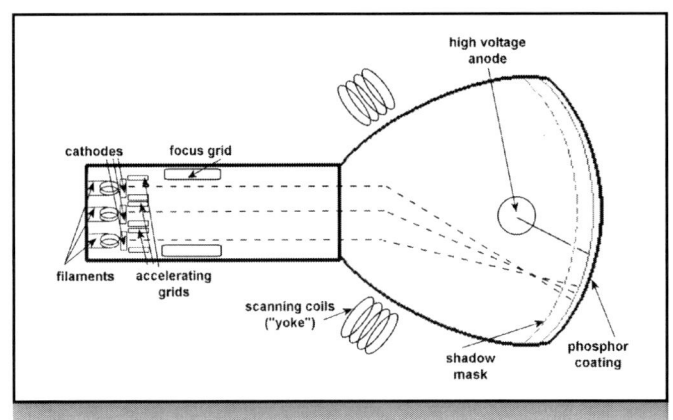

Chemical compounds (phosphors) that convert electron-beam energy into light of the additive primary colours (red, green, and blue), are deposited on the inner face of the glass picture tube in precision arrangements of stripes in alternating colours. If you look at the screen of a colour picture tube monitor through a magnifying glass, you can see that it's made up of many tiny line segment clusters that glow with red, green, and blue light.

At the rear of the picture tube is the electron gun, which produces three separate beams of electrons. These beams contain the picture information for the red, green, and blue colour channels. These hit the coloured stripe segments, and the tube is designed so that each beam can only hit segments of its own colour—a shadow mask prevents each beam from striking the others' stripes. Because the coloured segments are so small and so close together, the effect, when viewed from a distance, is of three superimposed images in the primary colours. Because the colour video signals adjust the strength of the three beams of electrons, the relative brightness of the image produced by each can be changed, and so the colour picture tube can display any colour in the additive colour system, as well as black, white, and all shades of grey.

With the advance of flat screen technologies, there are now very few colour picture tube displays being manufactured.

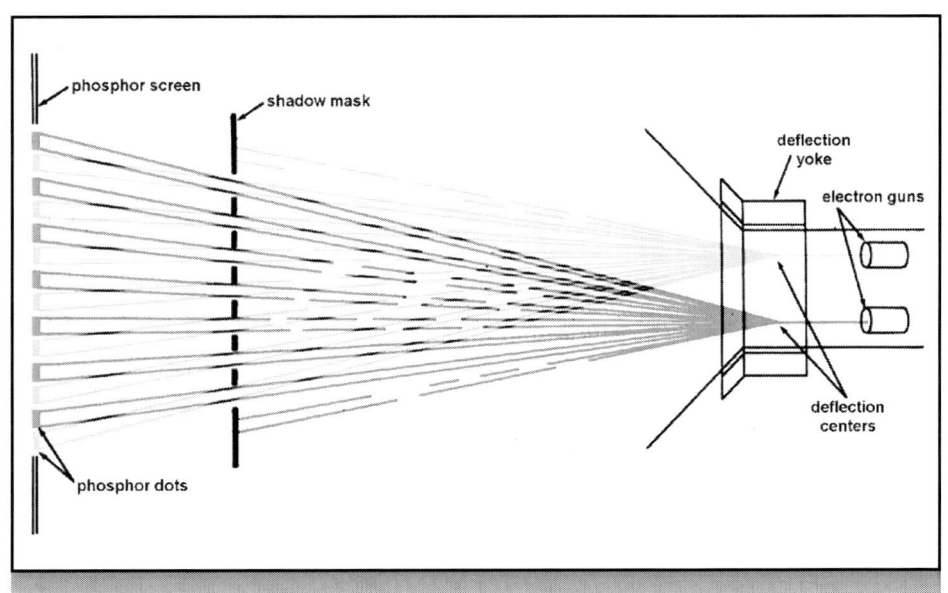

Typical colour television picture tube

Electron gun deflection and masking

FLAT SCREEN DISPLAYS

LCDs

One of the most common flat screen monitor types is the liquid crystal display (LCD). What started out as a small-size technology allowing the production of pocket-size portable television sets and laptop computer displays, has now matured into large screen HDTV quality. At one time, LCD displays didn't have the brightness, resolution, or contrast range of a picture tube, but improvements in these areas have resulted in LCD displays that are now being used as picture reference and evaluation monitors in the professional television industry.

Liquid crystals are neither completely solid, nor completely liquid. Clearly, the panel on which we view images is of solid construction, but if you press on an LCD display gently with a fingertip, you can see how easily it can change its state—you will notice colours are formed by the pressure from your finger. That's because you are twisting some of the crystals and as a result, they are letting different amounts of light through the panel.

Black-and-White LCDs

LCD panels are a sandwich of several thin layers. In a simple monochrome LCD display, light is created with a back light panel, made up of either fluorescent tubes or white light emitting diodes (LEDs) with a white diffusion screen in front of them to even out the illumination. This light, at the back of the panel, is first polarized by a filter. That means that the light wave will only pass through this filter if the wave is approaching it with a particular wave direction (for example, only completely vertical.) This polarized light is then passed through the liquid crystal layer, which, in a non-energized state, allows the light to continue on through, but twists the polarity of the light wave 90 degrees as it does so. The liquid crystal layer also has two electrode layers, one on each side of it, which will allow the display to energize certain portions of the crystal when required. Finally, the front layer of this panel is another polarizing filter, but this one is arranged so that its polarizing action is at 90 degrees from the back layer.

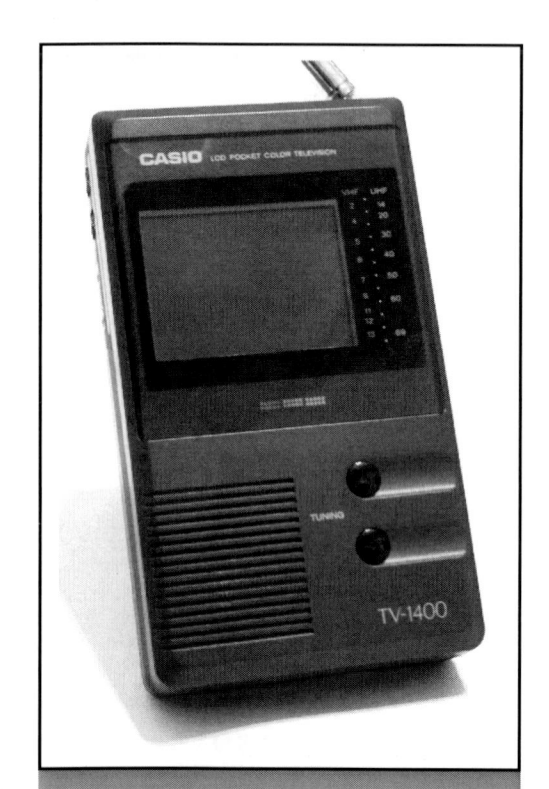

Early use of LCD displays—a portable television set

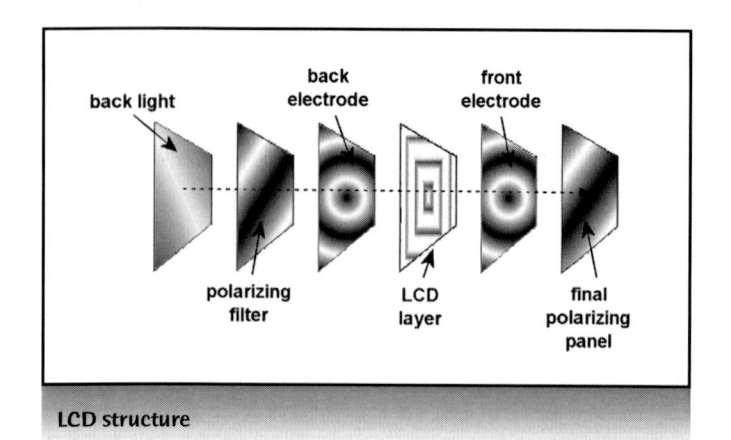

LCD structure

Therefore, when the LCD is de-energized, the light from the back of the display is polarized, then sent through the crystal where its polarity is twisted another 90 degrees, then sent to the front filter where it will get passed on to the viewer, relatively unobstructed. If, however, the crystal is then energized, it will untwist, and therefore not change the polarity of the light wave coming from the back. This means that the light will now be blocked by the front filter, resulting in a dark or black area on that portion of the screen. By varying the amount of electrical current on the crystal, the amount of untwisting can be varied, resulting in various shades of grey.

Colour LCD Displays

Inside a colour LCD display, clusters of LCD crystal are placed very close together, in rows and columns. A group of three of these subpixels have red, green, and blue colour filters over them. By varying the electric currents of each of these crystals in a given group, various amounts of red, green, and blue light can be emitted from the display, resulting in the creation of different colours. Because these subpixels are so close together, the final image doesn't appear to us as a cluster of individual light sources, but a mix of all three of them, similar to the effect in a CRT display.

One of the reasons LCD displays took so long to grow to large proportions is because each little pixel in an LCD display is turned on and off by a very small transistor. Historically, it was difficult to grow large arrays of transistors in anything other than very clean rooms, and even then, there were still flaws, which meant that the entire display had to be discarded, since it wasn't possible to repair a few separated, single pixels.

Close up of LCD screen, showing various colours' display

Gas Plasma Displays

In plasma displays, two channels of gas "discharge-tubes," etched into two plates of glass, are laid on top of one another, perpendicular to each other. These channels contain xenon and neon gas. The intersection of the two channel layers creates the picture elements. Each one of these subpixels, in turn, has electrodes imbedded into it, so every junction is digitally addressable. On top of this matrix are alternating red, green, and blue phosphors which will generate the colours. There is one vertical channel for each of the phosphor colours, which means that there are three times the number of channels as there are full pixels—a 1920 × 1080 display will have 5760 vertical channels. There is, of course, one horizontal

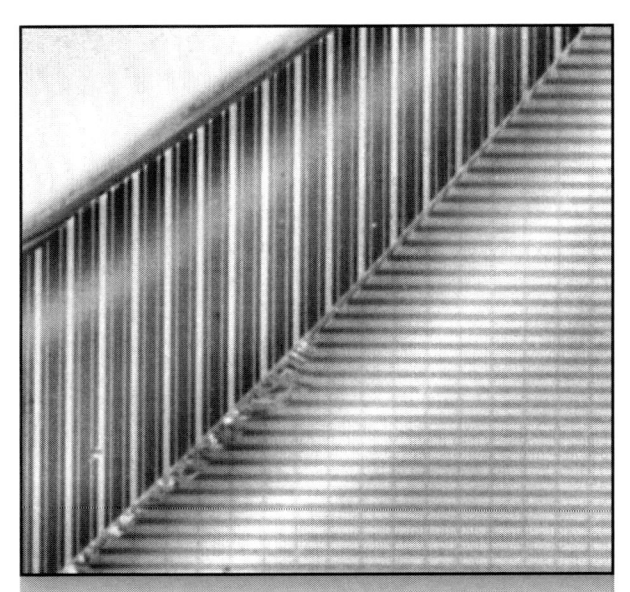

Cutaway of plasma screen, showing channels in glass plates

How gas plasma screens work

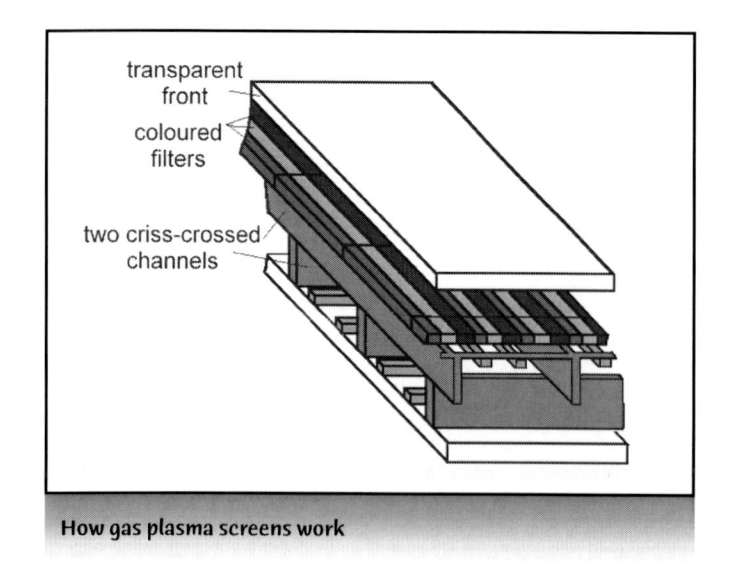

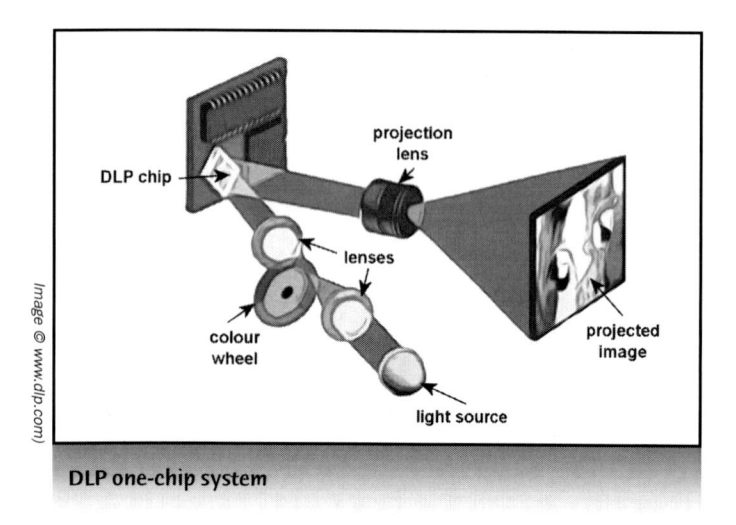

DLP one-chip system

channel for each horizontal line to be presented on the screen. When two crossed electrodes are activated with electric current, an ultraviolet light is generated within the gases. This invisible UV light is turned into visible light by one of the group of three closely spaced phosphors. Because the light is generated by the phosphors, no back light is needed in this system.

DLP Displays

Digital Light Processing, or DLP, displays are a relatively new technology that use an array of tiny mirrors, all separately addressable, fitting in a space less than two centimetres across. The mirrors reflect a high intensity light source and, by tilting 10° back and forth, they allow light to reach a projection screen. The mosaic of all these tilted and nontilted mirrors creates the picture on the screen. There have been systems developed using one, two, or three DLP devices.

The one- and two-chip systems use a colour wheel to produce the full spectrum—each colour channel's information is sent to the DLP in turn, in synchronization with the colour wheel's revolutions. In this way, the system projects red, green, and blue information once for each television field. If you wave your hand in front of this projection or move your eyes too quickly, you'll see the individual R, G, and B components on the projection screen.

The three-chip version uses a colour splitting prism similar to that used in a colour television camera, so, while the cost is greater, the "break-up" problem is no longer an issue.

Large Outdoor Display Devices

The most common system used in outdoor displays is an array of high-intensity illuminators (usually light emitting diodes, LEDs) which produces a very bright, high contrast, reasonable resolution display. These panels can frequently be seen in public places; outdoor advertising displays and rock concerts are two of the most frequent locations.

PROJECTION TELEVISION SYSTEMS

Larger television images can be achieved using a variety of projection systems that are designed for home or professional environments.

Front Projection

Larger screens normally employ front projection, using either LCDs or DLP techniques, and can produce television images on theatre-size screens.

LCDs

These projectors are becoming very popular with the maturity of high density liquid crystal displays. The LCDs are backlit with a strong metal halide projection lamp and, via a standard lens system, the images are projected on a screen.

ILA

ILAs, or Image Light Amplifier systems, use a special LCD panel with highly reflective back surface. Polarized light illuminates the panel, and it reflects back varying amounts of light, dependent on the information at a particular position in the image. Three of these systems are used, one for each of the primary red, green, and blue colours. The three images are combined through additional optics, and the final image is projected on the viewing screen.

DLP

DLP front projection systems use the same technology as the rear-screen DLP television monitors described earlier, except that the final projected image is sent through a typical projector lens system onto a distant screen.

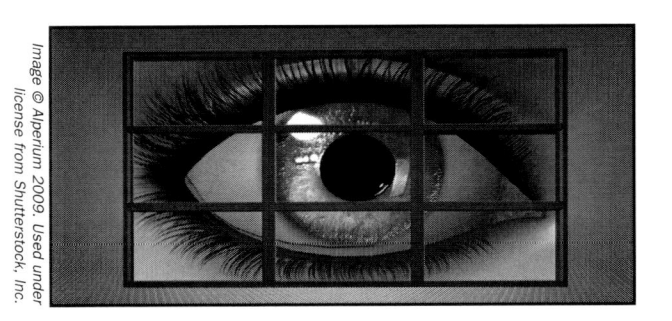

Videowalls

So-called "videowalls" are nothing more than several monitors arranged in a grid, each displaying only a portion of the complete television picture. Usually the displays are 25" to 27" RGB displays, although occasionally they can be up to 40", measured diagonally. Sophisticated computer control and digitizing hardware permit the proper part of the television picture to be displayed on the corresponding monitor.

How to Line Up a Colour Monitor

A colour monitor is a display device that sends detailed picture information to your eyes, which are very sensitive to changes in brightness, saturation, and hue. A properly aligned colour monitor can be as important a piece of test equipment as a waveform monitor or vectorscope.

But colour monitors can faithfully depict the quality of the pictures in a system only if the following conditions are met:

- the viewing conditions are correct
- the video system feeding it is correct
- the monitors have been correctly aligned

Philosophy of Television Reproduction

Both analog and digital video systems are designed to reproduce any scene as if it was illuminated by daylight, regardless of how it was actually lit. A white object in the scene, with a properly white-balanced camera, will emit a white light from a properly lined up colour monitor as 6500 degrees Kelvin (daylight colour temperature). Therefore, all monitors should be lined up to reproduce this "reference white."

Viewing Environments

In the home, the typical television set emits some light of its own, occupies a narrow segment of the viewer's field of vision, is illuminated by daylight and living room lamps in widely varying amounts, and sits in often colourful surroundings. In other words, it exists in a noncritical, but pleasant viewing arrangement. The home

Ideal viewing environment

viewer adjusts to gradual changes in familiar objects and colours on the screen as long as the rest of the scene is in colour context and will interpret a range of somewhat desaturated colours as more or less "white."

But we can't get away with this in professional television. An ideal viewing environment in industry should have us sitting a certain distance away from and at a certain angle to the screen. Typically, this is about four to six monitor screen heights away, and within a ±30 degree angle of the screen's perpendicular. Other specifications include having a "surround" (the area around the screen itself) with a colour temperature of 6500 degrees Kelvin (neutral white colour), and a very low incident illumination. This is why so often we see control room personnel working in a relatively dark environment. Even the room decor and furnishings factor into this: they should be neutral colours, preferably with a matte finish to minimize monitor light reflections. Some stipulations even go so far as to state that the actual lighting in the room should be at 6500 degrees Kelvin, although this is rarely implemented in practice.

Many people wonder why we don't use much less expensive consumer monitors in our professional environment. The reason is that professional monitors are carefully calibrated for colour temperature, brightness, and contrast. Consumer devices don't have the controls to adequately set these parameters to the exacting degree required in the television industry.

As well, there are presently no agreed-upon reference white standards for domestic television receivers and monitors. The two predominant reference white factory calibrations are 6500 K and 9300 K. Probably the average home receiver is set somewhere between these two values, but closer to the higher colour temperature, i.e., more of a bluish white compared to a professional colour monitor. Many consumer television sets today also come with various personal viewing settings, such as "cinema," "games," "sports," and so on, further obscuring the standards.

The problems that come with less than ideal viewing conditions are many. If the observer is seated farther away from the monitor than the recommended distances, too much of the surrounding area will be taken into account when judging the brightness and colour of the television picture. If seated too close, the viewer will eventually suffer from eyestrain.

If the surround is not of neutral colour, the monitors will appear to display the complementary colour. If you were to do fine colour correction on a video camera using a reference monitor in this non-neutral environment, it would be balanced to display a visual match with the area surrounding the monitor. Consequently, the colour balance of the shot for everyone else who viewed it would be altered, regardless of their viewing conditions. For example, if a professional bank of television monitors was flooded with pink light (for example, on a set, as a background for a shot), all the monitors within that bay would appear decidedly green. They would also probably be of lower contrast due to the amount of light spilled on them.

How to Line It Up

Techniques for Professional Monitors

Having shown why even a correctly aligned monitor will frequently not appear as it should, here's one recommended procedure for lining them up. It is suggested that professional colour monitors be checked and adjusted every few days, although some procedures outlined below can be done less often.

To do this lineup, you'll need two things:

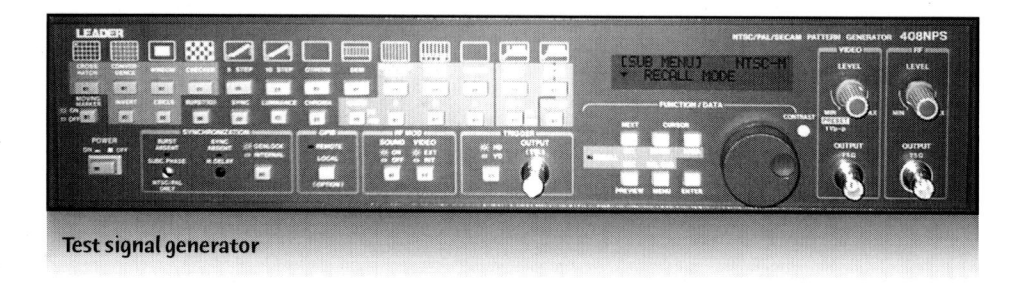

Test signal generator

1. some test signals (colour bars, grating and black, and maybe also a window and multiburst) that are routed to the monitor in the same fashion as the rest of the signals you'll be sending to it,

2. your eyes

Typical broadcast monitor lineup controls

To begin, ensure that the monitor has been on for at least a half hour (preferably an hour). The room ambient lighting should be the same as when the monitor is in normal use. All controls should be in their preset or centred positions.

One more point: all monitors should be clean. Be sure that all monitor faces have been wiped clean, and are not crusty with dust first; you may find your brightness, contrast, and colour problems can be solved with a little glass cleaner and some elbow grease.

Alignment procedures with an asterisk [*] below should be left to engineering, but all aspects of picture quality can be judged by operators to determine problems.

ALIGNMENTS APPLICABLE TO CRT MONITORING SYSTEMS

Purity

This is a problem unique to CRT monitors. A lack of purity manifests itself by colours that are not the same from one part of the picture to another, or colours that appear as blotchy. Purity shouldn't have to be checked more than once a month, unless stray magnetic forces have accidentally come near the CRT monitor.

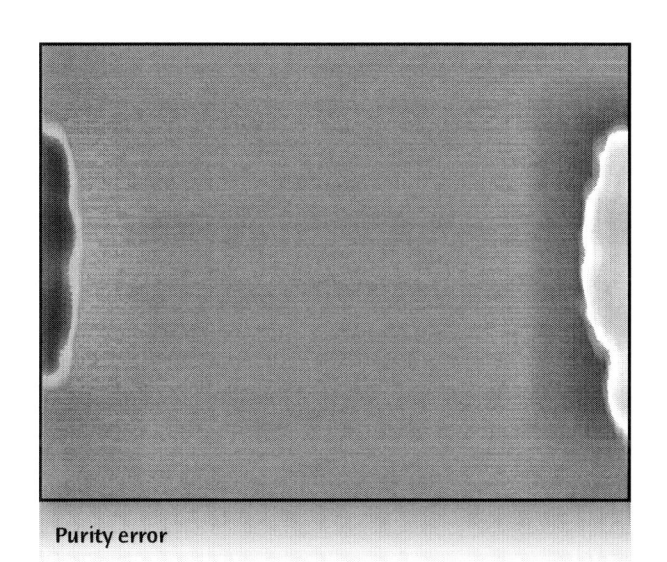

Purity error

There are two ways to check for possible purity problems. If you have a test generator, feed the monitor with a "red field" signal. This is simply a full screen of pure red colour. Alternately, if your monitor has the ability to do so, turn off the green and blue beams, leaving only the red, and turn the contrast control to minimum and increase the brightness until a medium bright red raster appears. Look for a uniform colour. If you find that the colour is varying from one area of the screen to another, use the degausser switch on the monitor, if it has one. If you still don't have good results, be certain that no stray magnetic fields are near the monitor, e.g., an audio speaker.

The internal degaussing system usually works fine. However, if you must use a degaussing coil, here's how. Remove your watch and other magnetically sensitive jewellery and adornments. Finger rings, earrings, and other body piercings can be left on, though. Hold the coil a metre or two away from the monitor and switch it on. Approach the monitor slowly, moving the coil in a slow circular motion parallel to the monitor face, until you are a couple of centimetres away. Back away from the monitor to your original starting position a couple of metres away, and turn off the coil while it is perpendicular to the picture tube.

Degaussing coil

It's clear from this discussion of purity, that you should never put your stereo speakers, refrigerator magnets, and similar magnetic things on the front face of (or even near) your home colour television set, if it uses a CRT as its display. You'll get purity problems for sure. What's worse, the purity issues won't go away even after you remove the magnet. If you have inadvertently messed up your set, you'll probably have to use a degaussing coil to fix it. Ask your local television engineer, and he or she may let you borrow theirs.

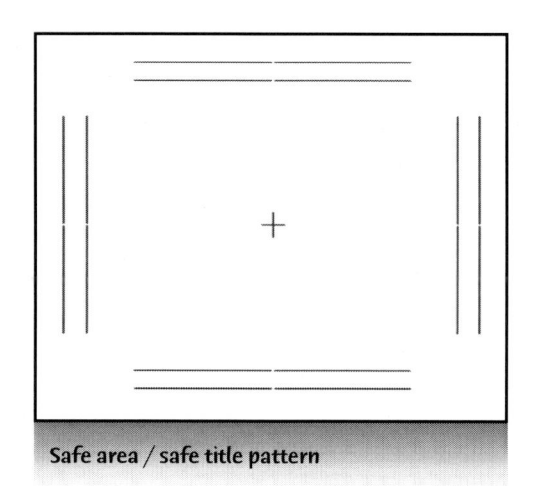

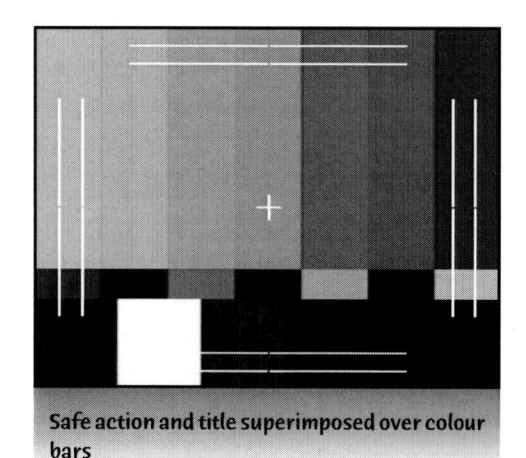

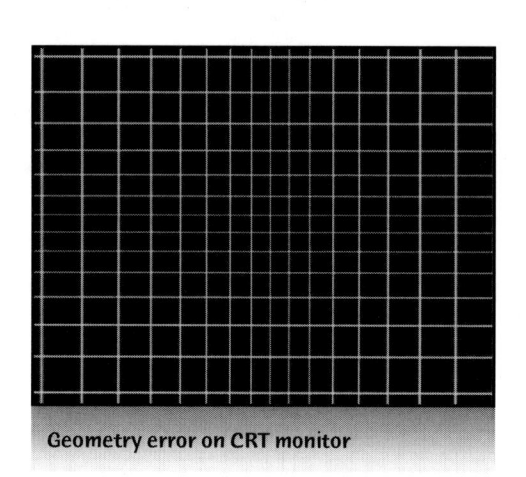

Scan Size *

Most consumer CRT colour television sets (and some other types of consumer television displays) don't display the entire picture that we originally shot with a camera. They enlarge the picture on the screen a little bit, so that a small percentage around the perimeter of the original framing isn't seen by the home viewer. This is known as "overscanning" the monitor.

In broadcast environments, our monitors have controls to allow us either to see our picture content the way the viewers at home do or to see all of our video right out to the edges. This control (often simply a switch) is called "overscan/underscan." In overscan, we see the video with some of the perimeter missing; in underscan, we see the entire picture content. Underscan should be used on all "source video" generators such as camera viewfinders and recording equipment where checking of picture impairments is vital. Overscan should be used elsewhere (including the program monitor).

Normal scan should be adjusted so the four corners of the picture are not visible and the safe action area is just visible. This represents an overscan of about 5%. If you don't have a test generator with a safe action/safe title generator, you can use SMPTE colour bars to adjust this approximately. With colour bars as the source, slightly more than one-half a bar width should be seen on either side of the screen.

Geometry and Aspect Ratio *

Geometry is the monitor's ability to scan evenly from left to right and top to bottom. This can be seen by using the test generator's standard grating pattern. The pin cushion and scan linearity controls should be adjusted so the picture appears without distortions when seen from normal viewing distances.

Focus *

This is electronic beam focus (the narrowness of the beam) and should be adjusted for best picture resolution in the brightest and central areas of the picture, without any noticeable loss of definition in the corners.

Safe area / safe title pattern

Safe action and title superimposed over colour bars

Geometry error on CRT monitor

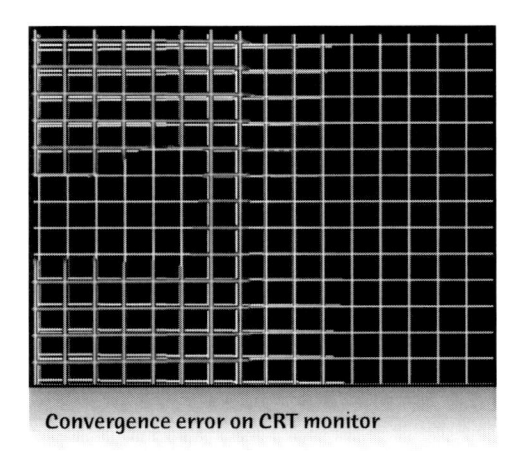

Convergence *

While looking at the standard grating signal, the white lines should not diverge into their separate colours anywhere on the screen. If they do, the convergence controls at the back of the CRT video monitor should be adjusted until there is no longer any colour fringing visible on the screen.

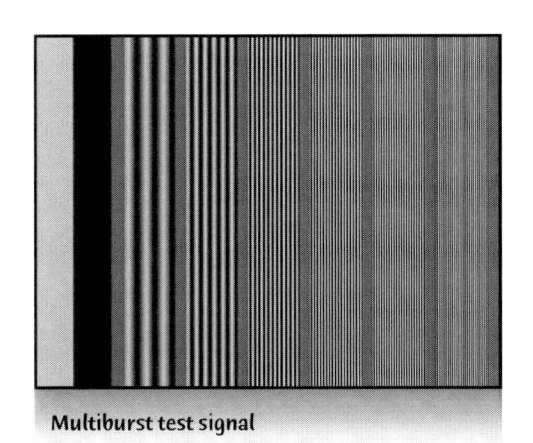

Convergence error on CRT monitor

Aperture Correction

This test requires a generator-produced multiburst test signal, not prerecorded and played back from a recording device. The aperture control corrects for high frequency losses due to cable length, and guarantees that the luminance section of the monitor has flat frequency response. This is rather hard to tell with the naked eye; the best method is to have engineering use an oscilloscope to check the luminance amplifier. The practised eye may be able to set this properly by watching the multiburst signal with the monitor in black-and-white mode.

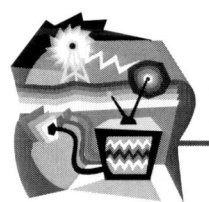

Multiburst test signal

ALIGNMENTS APPLICABLE TO ALL MONITORING SYSTEMS

Colour Saturation, Hue, Brightness, Colour Temperature, and Grey Scale Tracking

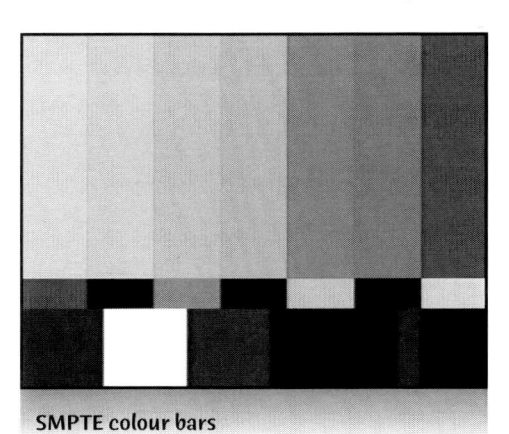

Saturation and Phase

With an accurate set of colour bars, and viewing only the blue channel, you should see four blue bars of equal brightness with black between them. These bars should increase and decrease in brightness equally as the contrast is reduced and increased over its normal range—they should "track together."

The left bar (the grey one in colour bars) is the reference; the hue control (phase) will affect the brightness of the inner two bars. The chroma control (saturation) will affect the brightness of the rightmost bar and the two inner ones.

SMPTE colour bars

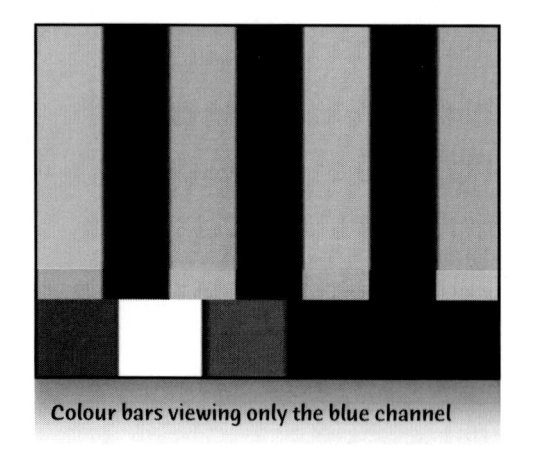

The sequence to follow is this: first adjust the chroma control, then the hue control so all four bars are equal with black in between. The four bars should now track together when you change the contrast control.

The above test is much easier with SMPTE colour bars—the ones with the complementary colours directly below the regular ones. With these bars, you can check brightness differences without the effects of purity or shading getting in the way. If you are using SMPTE colour bars, you can adjust the saturation and hue controls until the upper portions of the bars match their brightness with the corresponding complementary swatches underneath them.

Colour bars viewing only the blue channel

Brightness

Brightness is adjusted correctly when the active scanning lines are just at the point of visual extinction from your normal viewing distance. This is easy to set up with the PLUGE (Picture Line Up Generating Equipment), usually found in the lower right-hand corner of the black area in the SMPTE colour bar signal. It consists of two small vertical ribbons, one slightly greater than normal black and the other slightly less. The brightness control is adjusted so the darker patch just merges with the reference black, but the brighter patch is still clearly visible, from where the monitor will be normally viewed, and under the normal ambient lighting conditions. Since monitors are normally viewed from more than two feet away, be prepared to enlist a fellow operator, do a lot of walking back and forth . . . or grow longer arms.

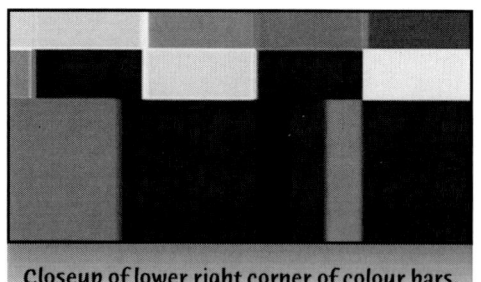

Closeup of lower right corner of colour bars, showing PLUGE

MONITOR MATCHING

What If You Have Several Monitors in One Place?

This is a common fault with colour monitors that are viewed next to each other—they never appear to look exactly alike. To compound the problem, monitors cannot be completely checked using a static test signal such as colour bars. They may match on bars but do not match when live action scenes change. This is never more noticeable than when you put a colour background generator in several monitors and amaze yourself at the various hues presented to you from just one pure generated colour. Cyan and magenta are particularly difficult to match.

This may happen for several reasons: different brands and models of monitors being placed next to each other, monitor sensitivity to light and dark scene changes, or incorrect colour decoding inside the monitor.

To match colours on more than one monitor, perform each of these steps with each of the monitors you're trying to match, in turn. That is, perform the first step on each monitor before moving on to the second step.

- Adjust for colour saturation and hue, brightness, and contrast as discussed earlier.

- Ensure you're in colour mode on all monitors, and readjust for any minor discrepancies with brightness (use PLUGE), phase (use the yellow and cyan bars), and saturation (use the red and magenta bars).

- Display moving pictures or a background generator from a production switcher. If they don't match, double-check your saturation and phase settings.

- When optimizing the monitors for a match, you have to realize when it's time to call it "close enough for television." You'll only get frustrated if you try for perfection. Let experience be your guide.

Black-and-White Monitors

Don't forget that black-and-white monitors need to be set up properly, too. You should set up contrast with full colour bars, and brightness with the PLUGE.

A Few Notes on Comb and Notch Filters

We've all seen the switches on professional monitors and some high-end consumer products: "Comb Filter," "Notch Filter," "Noise Reduction," and "Sharpness." Which switches should you turn on or off?

The Comb

The comb filter removes the colour information from the luminance information by taking the original video signal, and adding to it a slightly delayed version of that same signal. Often this uses two adjacent lines of video, which are then added together and sent to the display. This gives the appearance of a sharper picture. You'll also eliminate the problem effect known as "cross-colour," which appears as a rainbow pattern whenever there are closely spaced lines (the "colour fringes in the sportscaster's checked suit" syndrome). But if you turn on a comb filter, you'll get some NTSC artifacts like "dot crawl." You'll see this on any hard vertically-based chroma transitions. If you want to see some dot crawl, just look for anything in the picture that has chroma on one scan line and not the next, for instance, a red "underline" in graphics. You will likely see little "dots" moving sideways, most noticeably along the bottom edge.

So, you can either live with the colour fringes or put up with the dot crawl, depending which way you flip the comb filter switch.

The Notch

The notch filter does just what the name implies. Within the luminance information, it "notches" out a band of frequencies (specifically, those around 3.58 MHz colour subcarrier). This gets rid of the comb filter "dot crawl" mentioned above, but, like any filter, it isn't perfect; it also makes your picture softer, as you're removing some of the detail information in the luminance channel.

Deciding which setting to use for the notch filter switch is up to you. If the dots bother you, run with the notch "on." If you prefer more resolution, run with the notch "off."

Noise Reduction

Noise reduction has a somewhat more severe effect. The noise reduction removes low-level detail from the picture, which (the designers hope) is mostly noise. It also can lose some picture detail, and often leaves things looking smeary. Most broadcast professionals aren't too fond of noise reduction systems in monitors.

What to Flip?

With broadcast monitors, the notch filter can normally be left on. Because of the quality of the signals coming in, as well as the high-quality design of the monitor itself, the sharpness isn't affected all that much. In general, if you normally sit back a couple of metres or so from the set, you probably won't be bothered much by the comb filter.

At home, do what pleases you most. If you like lots of detail then crank up the sharpness and flip the switches that look good to you. Other people may prefer a less crawly, more subtle television picture, though.

How to Line It Up

Techniques For Consumer Television Sets

Find a station that's running that perennial favourite of late night programming, bars and tone. Now, look carefully at the colour bars and follow these directions.

- **Tint (or Hue):** The yellow bar should be a nice lemon yellow—no green or peach tones in it. The magenta bar shouldn't be purple or orange. Adjust until this is so.

- **Colour:** The bars shouldn't be blasting with colour, but they shouldn't be weak or pastel, either. Adjust to suit your eyes.

- **Brightness:** Look for a PLUGE signal in the lower right corner. If the station's bars don't have one, find another station. Adjust the brightness so the "lighter than black" PLUGE bar is visible, but the "darker than black" one is invisible.

- **Contrast:** This is a personal, do-it-by-eye, adjustment. Adjust the contrast not so high that things are blooming or smearing all over the place, but not so low to make things dull looking. You want bright, clear colour bars using this control.

- **Sharpness:** Don't make it too artificial looking, but a little edge on the bars is nice. You might want to switch to a late night episode of regular programming to really check this, because it's easier to see the effect on normal pictures.

9
CAMERAS

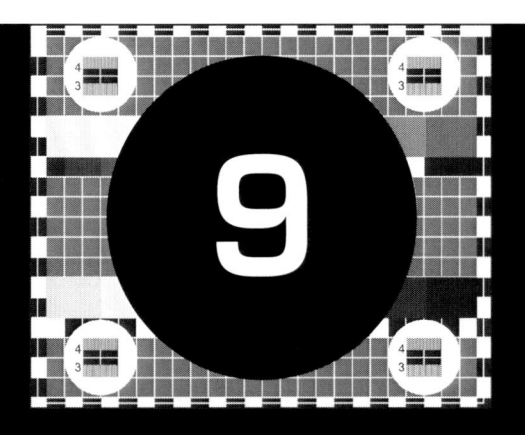

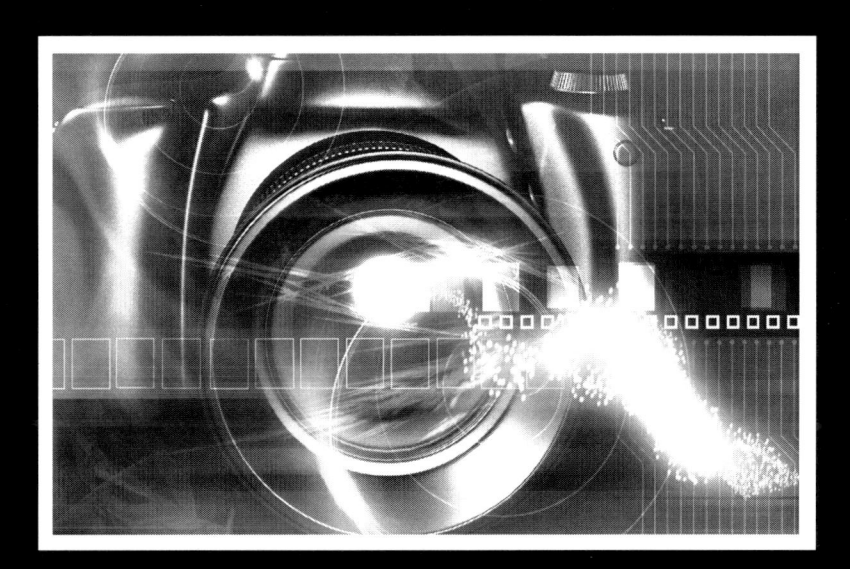

*"Time has convinced me of one thing.
Television is for appearing on, not looking at."*

—Noel Coward (attributed)

BLACK-AND-WHITE CAMERAS

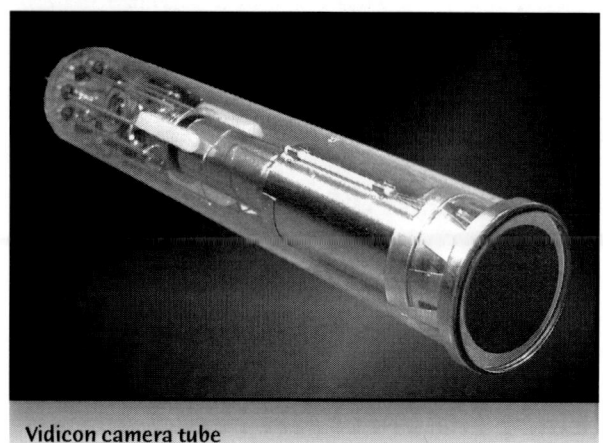

The Camera Imaging Device

The principal elements of a typical black-and-white television camera are the lens and the camera imaging device. From the beginning of electronic television to the 1980s, this would have been a camera tube (with its associated scanning and focusing coils). Now it is usually a charge coupled device (CCD). The camera lens focuses the scene on the imaging device.

CCDs

Their light weight, low cost, and high reliability allowed CCDs to gain rapid acceptance in the television industry. Manufacturers now produce these devices for use in professional and consumer video camcorders, as well as digital still-photo cameras.

How CCDs Work

There are four sections in a CCD: a layer of photo diodes, a capacitor, a transfer-gate layer, and an address decoder.

The topmost layer of a CCD is an array of photo diodes. As varying amounts of light strike the diodes, those that are illuminated become "forward biased," and a current flows that is proportional to the intensity of the light. The more light there is at a particular point in the scene, the more current will flow to the next stage.

The current from each diode is sent to a solid-state capacitor in the CCD. Each capacitor stores the current sent from each of the photo diodes, and this happens over a very brief period of time. This is similar to a film camera shutter's opening for a brief instant to let light pass through to the film emulsion, and then closing and staying closed until the film is advanced and the next photo is taken. The variable "shutter" control on a CCD video camera allows us to adjust the duration of this high-speed video transfer.

The CCD charge transfer-gate layer, or analog shift register, deals with the charges stored in the capacitors. A useful metaphor is to think of the capacitors and charge transfer-gates as a high-rise apartment building—the capacitors are the apartments in the building, holding their charge contents. The transfer gates are the apartment number addresses.

An address decoder cycles through the addresses of the registers, one after another, and the charge transfer-gate, using each address as location finder, reads out the analog voltages from the capacitors for each pixel. The speed of operation of this decoder is synchronized to the scan rate of television. To continue the apartment comparison, imagine that we knock on every door of every apartment, in sequence, and see what's inside: "Apartment 101, show me your contents. Apartment 102, show me yours . . ." As we move from one apartment to the next, and then from one floor to the next, and so on, we would eventually sequentially scan the contents of all of the apartments. The serial reading of all of the CCD chip's individual storage elements, one line after another, automatically creates the scanning pattern for the television image.

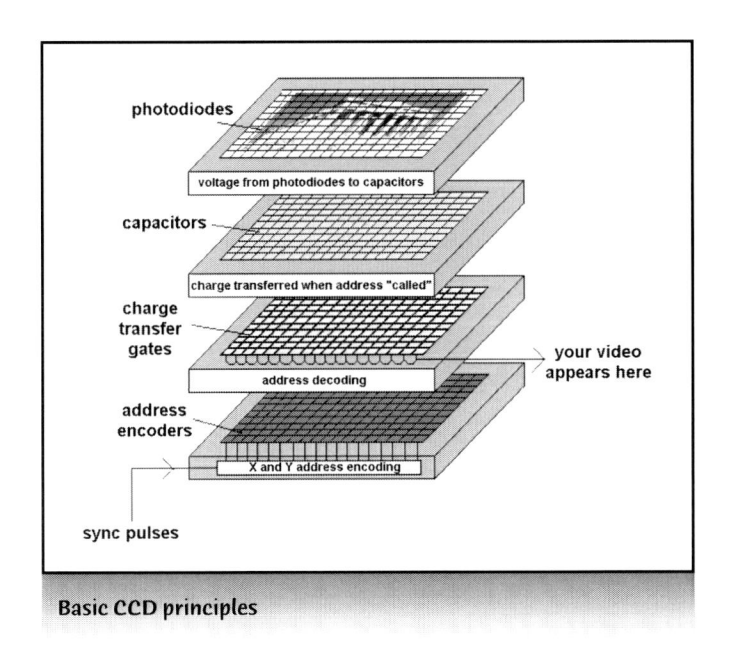

Basic CCD principles

The actual transfer of the voltages out as analog video is why CCDs are so ingenious. The CCD chip can transfer the voltage from cell to cell without any loss. This process is called charge coupling, which is how the CCD gets its name.

When the transfer-gate of a CCD is activated, the CCD's clocking circuitry moves the current stored in each capacitor picture cell to the adjacent cell. Clocking the shift registers in this manner transfers the video to the output, one pixel value at a time. The last cell in the chain sends its voltage, in turn, to the output circuit of the chip. As an added bonus, cycling through all of the cells this way will not only send out all of the stored voltages, but also discharges all of the cells, too. Everything goes back to normal and the cells are ready to take in new analog voltage values, for the next frame of video.

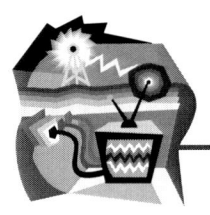

COLOUR CAMERAS

Three-Chip Cameras

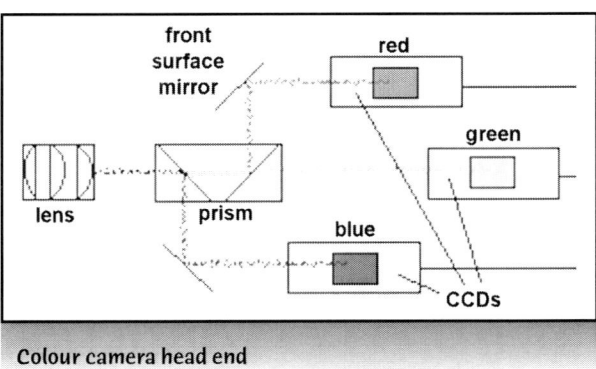

The three electrical signals that control the respective beams in the picture monitor are produced in the colour television camera by three CCD chips. The camera has a single lens, behind which a prism or a set of dichroic mirrors produces three images of the scene. These are focused on the three CCDs, which are rigidly and permanently mounted on the prism itself. In front of each CCD is a colour filter; the filters pass only the red, green, or blue components of

Colour camera head end

the light in the scene to the chips – one colour per chip. The three signals produced by the camera are transmitted (via colour encoding) to the video output of the camera, and eventually to the picture monitor, where the scene is re-created.

One-Chip Cameras

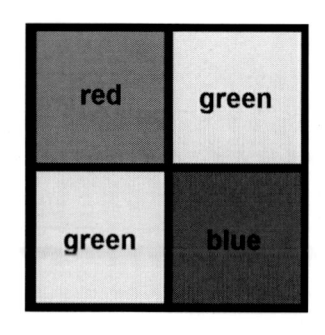

While three-CCD video cameras are now the norm, there are still one-chip systems in use, especially in digital still-photography cameras. A one-chip CCD camera can be produced using one of two methods.

The first uses red, green, and blue filters, or microlenses, over alternating pixels of a single CCD. Each tile of this mosaic is about one square micron in size (one-thousandth of a millimetre). Keep in mind that this will let in only about a third of the available light, since the filters block so much of the original scene's illumination. The camera's image processor knows which pixel corresponds to which colour while the CCD's currents are being read from the chip.

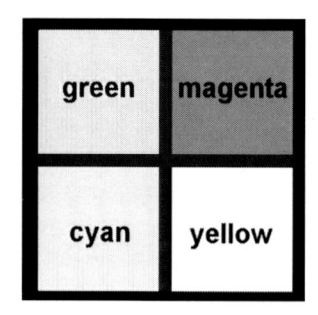

In other single-chip cameras, cyan, yellow, magenta, and green filters are used instead of RGB filters. The first two pairs of colour lenses (magenta + yellow, and green + cyan) decode luminance and the R-Y colour difference signal. The second two pairs (green + yellow, and magenta + cyan) decode luminance and the B-Y colour signal. This system works because the analog signals stored under each of the eight microlens elements represent various combinations of coloured light, which can be combined to create a single picture element of full colour video.

The second system described here will make the camera twice as sensitive as the RGB one-chip CCD method, but the colour reproduction at low light levels isn't as good, because the RGB information is derived arithmetically—the subtraction of two small numbers with lots of noise results in poor accuracy and a high noise level. This can be seen when looking closely at photographs taken with a one-chip digital still camera: there is often a lot of colour noise in the darker areas of the picture.

The colour mosaic method is limited by the ability of CCD manufacturers to make these specialized microlenses and position them accurately over a high-resolution CCD pickup device. As a result, the output of these one-chip systems, while of remarkably high quality, can't really compete with true professional HD video quality.

LENSES

An optical lens is a device that forms an image by the refraction (bending) of light rays. The "burning glass," for concentrating the sun's rays, has been known since ancient times. The magnifying property of a simple lens was first recorded by Roger Bacon in the thirteenth century.

A simple lens consists of a single piece of glass or other transparent material having two opposing faces, at least one of which is curved. A lens is said to be con-

convex/converging lens with real image

concave/diverging lens with virtual image

simple zoom lens

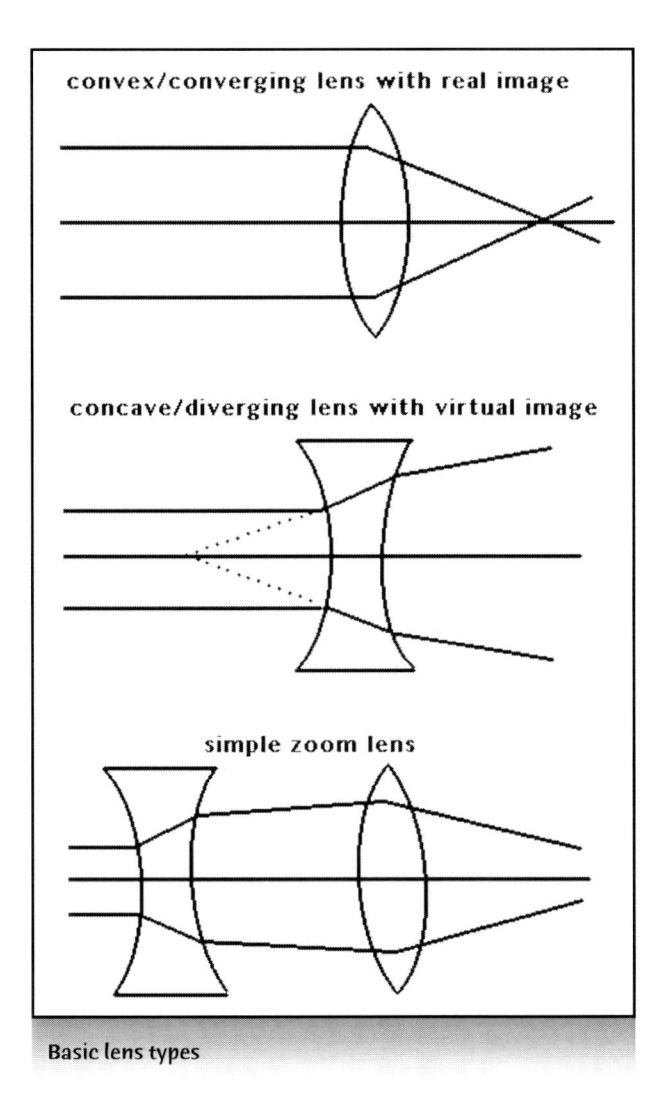

Basic lens types

verging, or positive, if light rays passing through it are deflected inward, and diverging, or negative, if the rays are spread out. Converging lenses are thicker at the middle than at the outer portion, while diverging lenses are thicker toward the edges. The surface of a lens can be concave (curved inward), plane (flat), or convex (curved outward).

The focal length of a single lens is the distance from the lens to the point at which incoming parallel rays focus. In other words, when an object is at an infinite distance away, and you can focus the real image on a screen, the distance from the centre of the lens to the screen is its focal length. Infinity, for sake of argument when dealing with practical camera lenses, is approximately one hundred metres away. In a negative lens, rays do not come to a real focus, but appear to start from a point called the virtual focus. The focal length of a diverging lens is considered "negative."

When we combine multiple lens elements, we have created a very simple zoom lens that can vary its focal length by changing the distance between the lens elements. The zoom lens was invented in 1956 by Pierre Angénieux. Before that time, in order to change focal lengths, the camera operator had to "rack" from one lens to another, available from a selection on a lens turret.

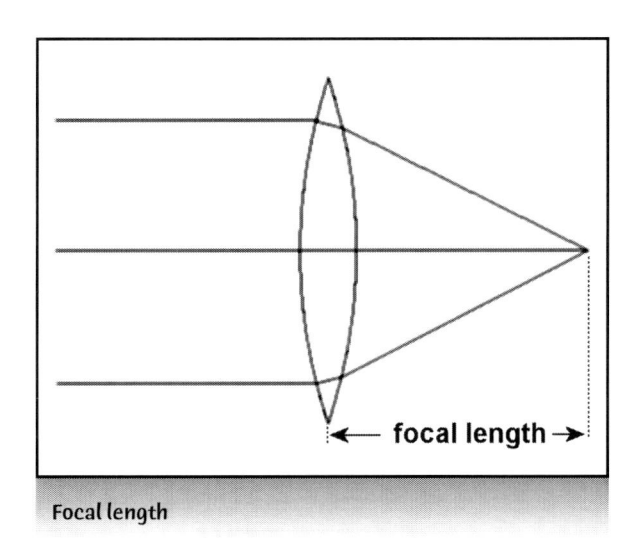

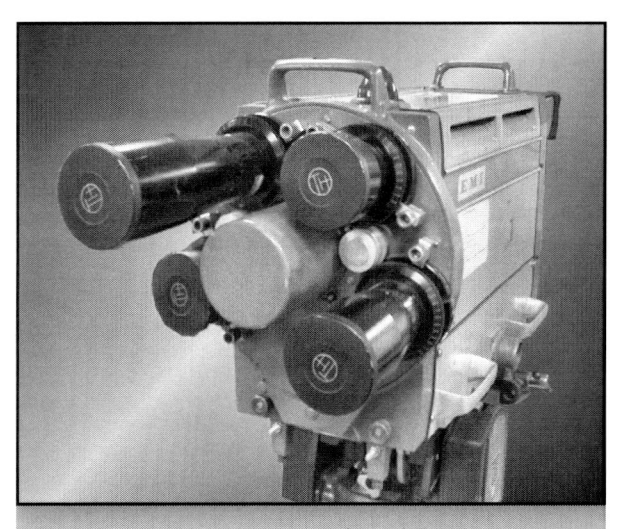

← focal length →

Focal length

Turret lens on television camera from the 1960s

Real and virtual images

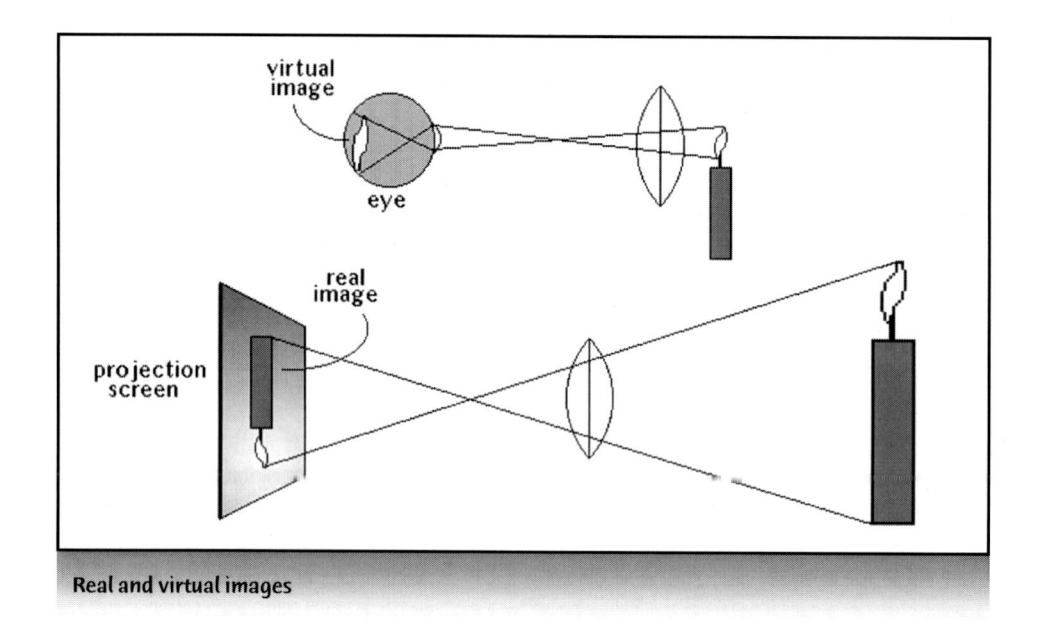

A simple convex lens used as a magnifying glass will produce a virtual image if the object is either at the focal point or between the focal point and the lens. If you look through a magnifying glass in this case, light from the object passes through the lens, and your eye focuses it onto the retina. The result is that you can see an image of an object that appears enlarged to you, behind the lens. That's the virtual image.

If the object is beyond the focal point of the same lens, a real image can be formed. A magnifying glass held in front of a candle will cause an image of the flame to be projected onto a screen or wall across the room. Think of a real image as one that you can actually reach out and touch. This is how lenses are used in a camera—a real image is focused onto the CCD.

Back Focus

As a camera lens zooms in and out, several lens elements are moving together to change the effective focal length. It's important that this varying focal length create a sharp image at all times throughout its zoom travel. Because professional video cameras can have interchangeable lenses, a back focus adjustment is used to ensure high-quality, crisp images.

How to Adjust Back Focus

1. Before starting, put the camera on a tripod and adjust your camera's viewfinder so it is in sharp focus. Ideally, you'd want the test pattern chart shown (which looks a bit like a dartboard) to be at least 20 metres from the camera, or as far as possible. If you don't have a test chart, use a page of printed text from a magazine.

2. Set the iris to manual.

3. Set the zoom to manual.

4. Open the iris to f/1.4 or its widest aperture. If the illumination on the test chart is too bright for the open iris, reduce the light or move the chart to a darker area.

5. Zoom the lens to its extreme telephoto position.

6. Focus on the chart.

7. Zoom out to a fully wide angle.

8. Loosen the back focus ring retaining knob.

9. Adjust the back focus ring for the sharpest focus.

10. Repeat steps 5 through 9 until focus is consistently sharp throughout the zoom.

11. When it is, tighten the back focus ring retaining knob to secure the ring.

F/ Stops

The f-number of a lens is the ratio of the focal length to the lens diameter. Lenses of large diameter have small f-numbers and hence greater light-gathering power than lenses of small diameter or large f-numbers. As an example, if we have a lens

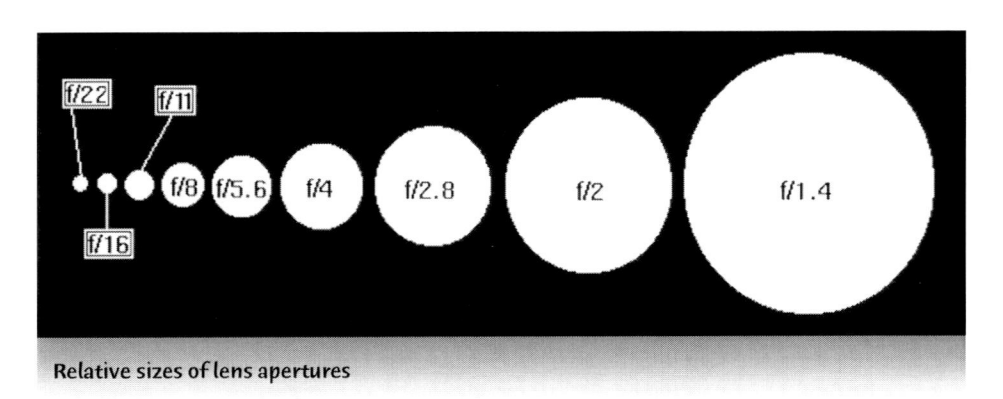

Relative sizes of lens apertures

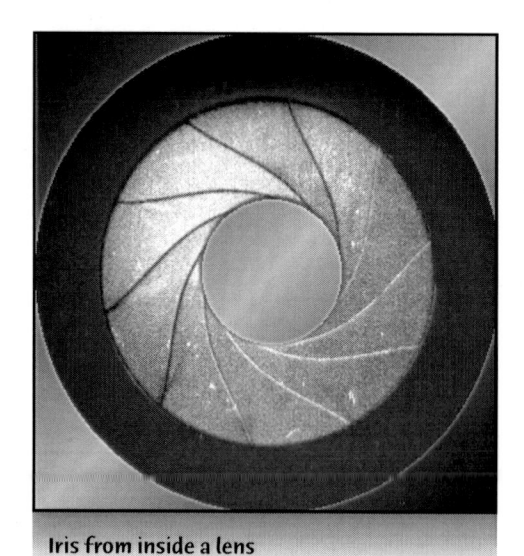

with a focal length of 50mm with a maximum iris opening of 25mm, it is considered an "f/2" lens.

We come across f-numbers when we speak of the iris of the lens, and they are usually expressed as a series of numbers on the iris ring. For example, if a 50mm lens is opened to a 12mm iris hole, it is set at f/4.

F-numbers (or f-stops) look strange: f/1.4, f/2, f/2.8, f/4, f/5.6, f/8, f/11, f/16 and so on. You'll notice that these numbers don't double each time. What they represent are the diameter changes of the iris setting, relative to one another. For each change by a factor of 1.41 (the square root of 2) in the diameter of the iris opening, the area of the opening doubles. This explains the unusual numbers. Remember that the amount of light passed by the lens is doubled each time a lens iris is adjusted to the next lower number.

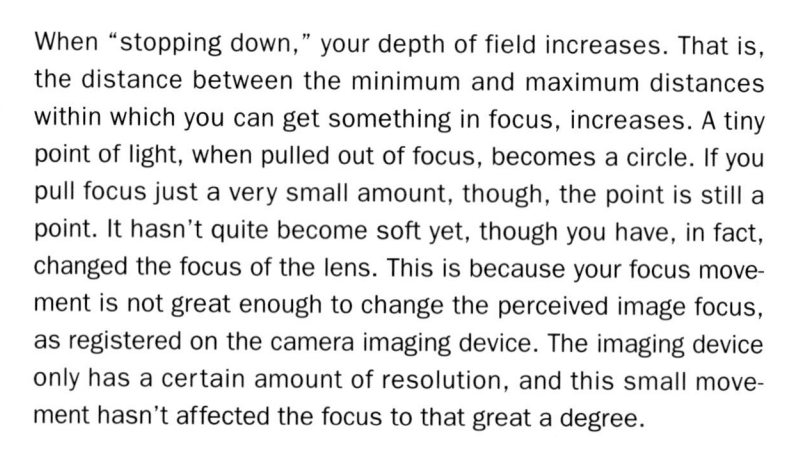

When "stopping down," your depth of field increases. That is, the distance between the minimum and maximum distances within which you can get something in focus, increases. A tiny point of light, when pulled out of focus, becomes a circle. If you pull focus just a very small amount, though, the point is still a point. It hasn't quite become soft yet, though you have, in fact, changed the focus of the lens. This is because your focus movement is not great enough to change the perceived image focus, as registered on the camera imaging device. The imaging device only has a certain amount of resolution, and this small movement hasn't affected the focus to that great a degree.

If you close the iris, you narrow the area through which light rays pass through the lens and therefore allow only the more parallel

Iris from inside a lens

F/stop numbers on a camera lens

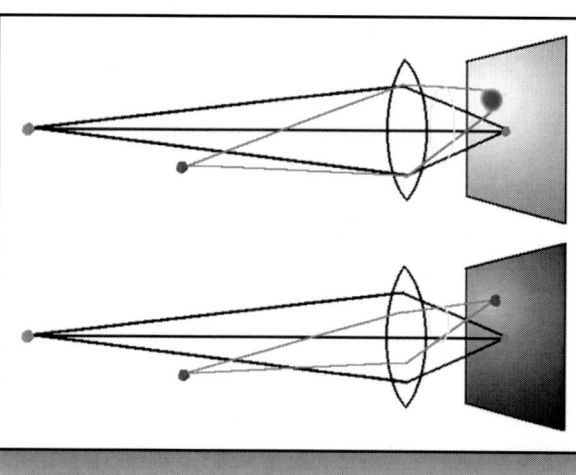

With a lens iris fully open, you get the most light, but the least depth of field - closer objects will be out of focus.

With a lens iris closed down, you let in less light, so your shot will be darker, but you get more depth of field - close objects will be in focus, as well as distant ones.

'Stopping down' increases your depth of field—more subjects, over a broader range of distances, will be in focus

⭕	**Larger** Iris	⚬	Smaller **Iris**
	smaller f/ stop "number"		**larger** f/ stop "number"
	lets in more light		lets in less light
	depth of field decreases - fewer things in focus from foreground to background		depth of field increases - more things in focus from foreground to background
Note: a Neutral Density filter allows you to open up to a larger iris (smaller f/ stop)			

rays. These are the ones passing through the centre of the lens (which bends light the least). Objects at greater distances from each other are able to remain within this range of camera resolution. You don't get something for nothing, though—you lose some of your light to increase your depth of field.

Depth of field is determined by several things:

- **the iris setting**—the more open the iris, the less depth of field

- **zooming in or out**—lenses set at maximum zoom have less depth of field than wide angle settings

- **neutral density filters**—these cut down light, therefore the iris will have to be opened more to compensate

- **shutter setting**—the higher the shutter speed, the shorter the exposure time, therefore the iris will have to be opened

OPERATIONAL PRINCIPLES AND FEATURES OF BROADCAST VIDEO CAMERAS

Most video cameras will have the features listed below. Each camera is different, in terms of types of controls used to adjust these elements, and their placement on the camera body. Sometimes you will find that some of these adjustments are dedicated switches or knobs on the camera; in many newer video cameras, these adjustments can be made by selecting various menu options. The photos that follow are typical; you may find that your controls look different. The best way to understand these controls is to read the owner's manual of your camera, and practice using the various functions.

White Balance Selector

This switch has such positions as MEMO (automatically adjusts the white balance, and stores it in internal memory); PRESET (a factory setting, sometimes can be overridden by the camera operator); and AUTO TRACK (the camera attempts, with varying degrees of success, to keep a check on the colour balance of the scene and correct for it "on the fly").

Auto White/Black Balance

Cameras should be black and white balanced as you move from one lighting colour temperature to another, for example, when shooting outdoors in daylight, then indoors later on using tungsten EFP (electronic field production) lights.

Perform the black balance process first. This will ensure that the black information in your shot has no undesirable colours. Then, while shooting a white subject (for example a piece of white paper), perform a white balance, ensuring that your camera is colour balanced for the type of light under which you're shooting.

Gain Switch

This two- or three-position switch allows you to gain more signal strength when working under very low light conditions. Although the gain allows you to work in a low light environment with little effect on the colours of the scene, it causes the pictures to become progressively noisy and soft because of the increase in video amplifier noise.

White Clip

All cameras have white clip circuits to prevent the output signals from exceeding a practical video level, even if highlights appear in the picture. Video won't go over a preset limit (usually 110 to 120 units) even if the scene has such highlights in it.

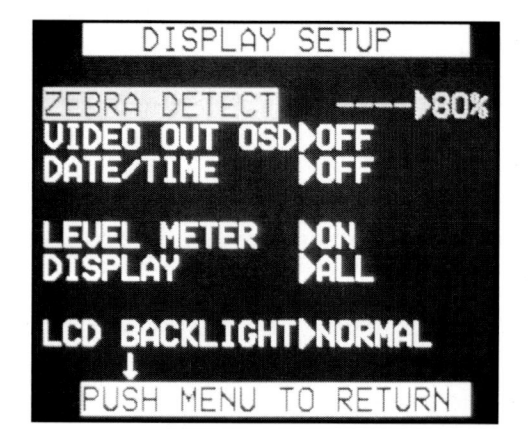

Zebra Stripe Level Indicator

This generates a "zebra stripe" pattern in the viewfinder by which you can judge whether your video levels are set for 70, 80, or 100 units (or more). These stripes are not recorded in the camera; they only appear in the viewfinder. Any picture elements above the peak level will have the stripes superimposed on them. Some videographers like to set their zebra levels to display 100 units of video, to check peak white levels. Others prefer 70 or

80 units of video, to check the exposure on skin tones. Set the iris so that the zebra stripes are only showing on facial highlights. This works for both dark- and light-skinned individuals, although the pattern will be larger and more pronounced on the latter.

Filter Wheel

Cameras have various neutral density (ND) and colour correction filters mounted on a filter wheel. Many of these wheels also have a "cap" position as well. The colour correction filters are used to reduce the blue colour temperature of sunlight, since ENG cameras are nominally balanced for tungsten lighting (with a colour temperature of 3200 degrees Kelvin). Some correction filters also have an ND factor added, to reduce the light level when shooting outdoors.

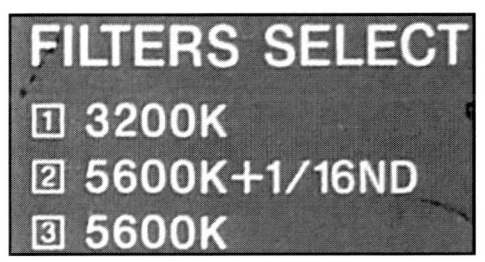

Shutter

CCD technology allows us to have action-freezing shutters built into our cameras. When using shutters, keep in mind that a higher shutter speed means less light is available, so the iris will have to be wider, or you'll have to put more light on the scene. Shutter speeds can vary from "off" (no shutter activation) to 1/8000 of a second. These higher speeds are great for sporting events featuring blur-free slow motion playbacks.

To experiment with the idea of a shutter, try this. Take a piece of cardboard and cut a long slot in it (how big doesn't really matter, but big enough that you can see through it.) Turn on an electric fan, running at low speed. Wave the cardboard back and forth quickly while looking at the fan blades through the slot. Experiment with different "wave" rates. You'll notice that the fan blades' rotating action appears to be "frozen" and is no longer blurry. This is the principle behind a television camera's shutter—the scene is exposed for only a brief moment in the camera, freezing the action.

Detail Level

Image enhancement is used in all cameras to improve picture quality. This is accomplished by raising the contrast between the light-to-dark and dark-to-light areas of a picture, by "overshooting" the electronic video signal at these transition points. The detail correction control allows you to vary by how much this will occur.

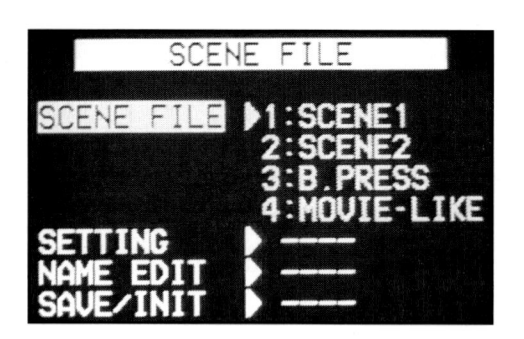

Scene File

A scene file is a small record of settings information that can be saved on a chip within the camera. Parameters that can be saved include white balance, iris and black levels, gamma, detail level, camera gain, iris settings, and other information. The memorized values can be recalled by pressing the corresponding scene file button on the camera or CCU (camera control unit). This allows the camera operator instant adjustment when moving from one shooting condition to another. As well, it allows multiple cameras shooting in one location to be normalized with identical settings—the scene file information for one camera can be transferred, using the chip, to the other cameras.

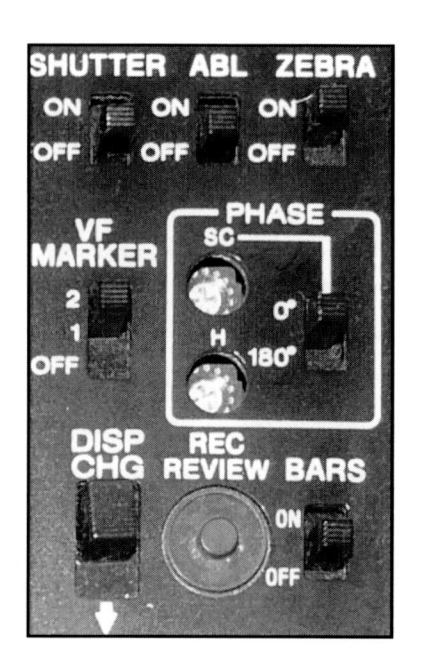

Bars Switch

The bars switch allows you to switch either from the picture output or colour bars. You should always place bars at the beginning of each recording when shooting in remote situations, so that the levels you record can be properly reproduced in the edit suite or wherever the recording is played back.

Viewfinder Controls

All viewfinders have the ability to be adjusted for brightness and contrast. Colour viewfinders can also be adjusted for hue and for colour saturation. Sometimes you will find these controls on the front of the viewfinder itself (especially true of black-and-white viewfinders), but now more commonly these adjustments are available through a selection on the camera's menu system. Remember that these adjustments only affect your viewfinder; they don't adjust the video quality of the actual recorded image. Make viewfinder adjustments using the internal colour bars generator, and following the procedure outlined in the Monitors and Television Sets chapter of this book.

Remote Control CCU (Camera Control Unit)

This is a remote control panel for a camera, and it has on it many of the switches found on the main camera body, and variations on controls found on ENG/EFP cameras (electronic news gathering/electronic field production cameras). The following controls are common, although others are also available:

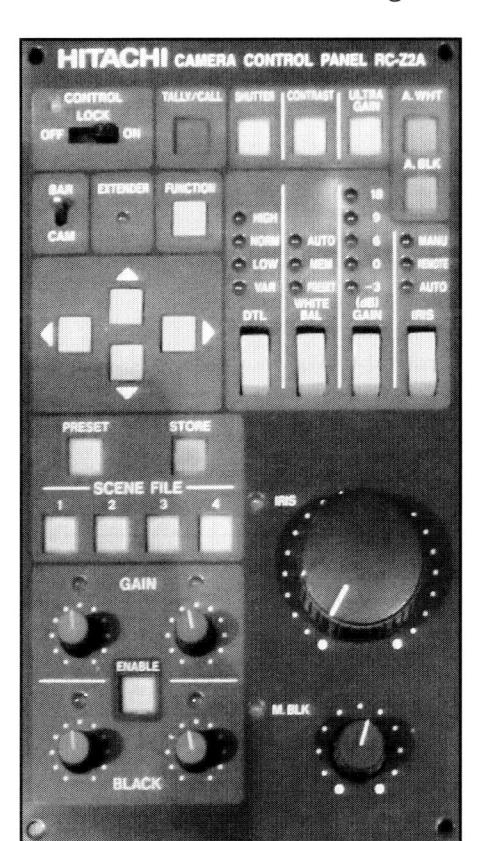

- control of your camera via the remote panel or the base CCU
- colour bars
- high sensitivity gains of 0 dB (no gain), and various increases up to +18 dB
- auto white and auto black balance
- white balance: factory preset (3200 degrees K); auto white setting; or manual adjustment via red and blue white-level controls
- black balance: auto black setting or manual adjustment via red and blue black-level controls
- master iris and black control
- iris setting modes: automatic control; preset (limits master iris control to 4 f/stops); or full (which does not limit master iris control at all)

Time Code Selection

Many cameras will allow you to preset the time code with push buttons, along with entering additional information in "user bits" of the code and selecting whether the code generator will run only when you press record (REC RUN), or run all the time, like a normal "time of day" clock (FREE RUN). Some cameras have a time code input (TC IN) connector to allow connection of an external time code generator, the code of which will be recorded in your camera.

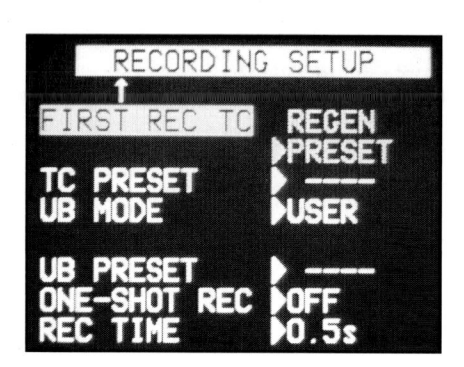

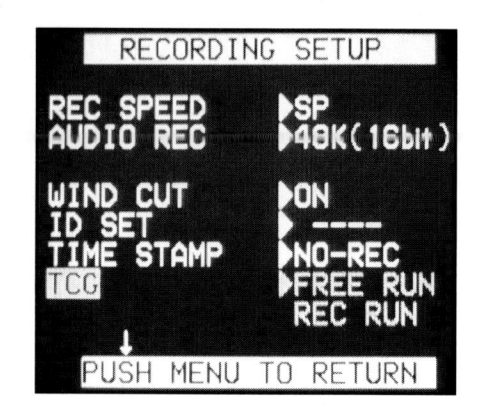

Video Connectors

Almost all cameras have a VIDEO OUT connector, to allow you to connect an external video monitor, so others can see what's in the camera's viewfinder for monitoring and checking purposes. In addition, there may be a connection for locking the camera's scanning to an external sync generator (GENLOCK IN).

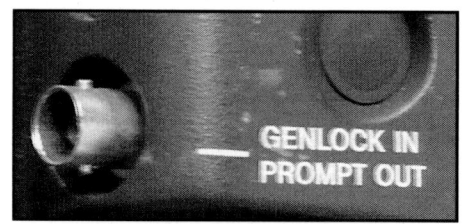

Audio Connectors

All professional video cameras will have XLR broadcast audio input connectors, one for each of the two available channels. Some cameras may have an XLR AUDIO OUT for monitoring playback within the camera. Every camera will have a headset jack which accommodates a 1/8-inch standard headset plug. When shooting, camera audio should always be monitored with a headset. Professional headphones are best, but a pair of earbuds will also work.

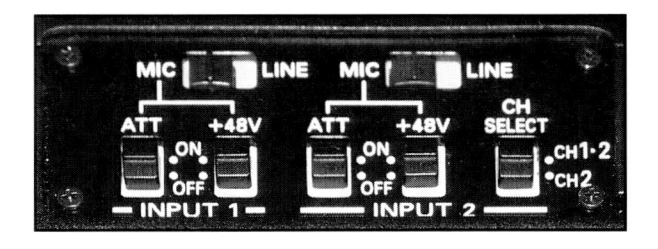

Audio Level and Input Controls

Many cameras have audio input level pots; you should adjust them as required for the optimum audio levels. There will usually be a switch to select which type of level (mic or line) is being sent to the camcorder recording system. Most cameras have an on-board microphone, which is often marginally acceptable, since it picks up the camera's zoom motor noise, and the sound of the operator moving the camera around and adjusting things on the camera body.

SPECIALTY CAMERAS

Specialized cameras are adaptations of existing camera technology put to new uses.

Stereo Cameras

Now that HD television is fully implemented, the next generation of video development will arguably be in the field of 3D video. Some form of 3D imaging has been around for decades—the motion picture industry has been experimenting with it, on and off, since the 1950s. Various developers are working on several different three-dimensional systems. Many have one thing in common: two identical video cameras placed side by side, recording two images simultaneously which are later played back through two projectors onto a screen. The two cameras cover what you would normally see through your left and right eyes. Other technologies involve using only one projector on the screen with various forms of dual-view integration. The days of red and blue 3D glass lenses are gone, having been replaced by polarized lenses that pass the left and right images to our eyes. Recent experimentation has also been done with 3D projection systems that do not require glasses at all. These are in their nascent stage as of this writing, but they have shown great potential for future development.

Multi-image Technology

This technology is usually implemented with a special system that has several cameras mounted on a sphere. Picture stitching software combines the images together into one panoramic view, which includes not only left to right pans, but also vertical views as well. The most common example of this is Google Street View. These images are created with a system developed by Immersion Media. While Street View is a still-image implementation, Immersion Media has developed a real-time video version as well.

Camera Mountings

Cameras can rest on many different objects: tripods, pedestals, your body (with or without a mount), sandbags, cranes, and even helicopters.

Tripods

A tripod is a group of three legs connected with various fastenings to raise or lower the overall height of the camera head. Sometimes a dolly is fastened to the bottom—a triple-armed unit with a wheel on each arm, to allow fluid motion around a studio or other smooth surfaced environment. At other times a simple "spreader" or "spider" is used at the base of the three tripod legs to keep them stable and to prevent them from sliding out from underneath the camera.

Pedestals

With the studio pedestal you can truck or dolly very smoothly, and elevate or lower the camera easily while on the air, by releasing the pedestal column locking ring or handle.

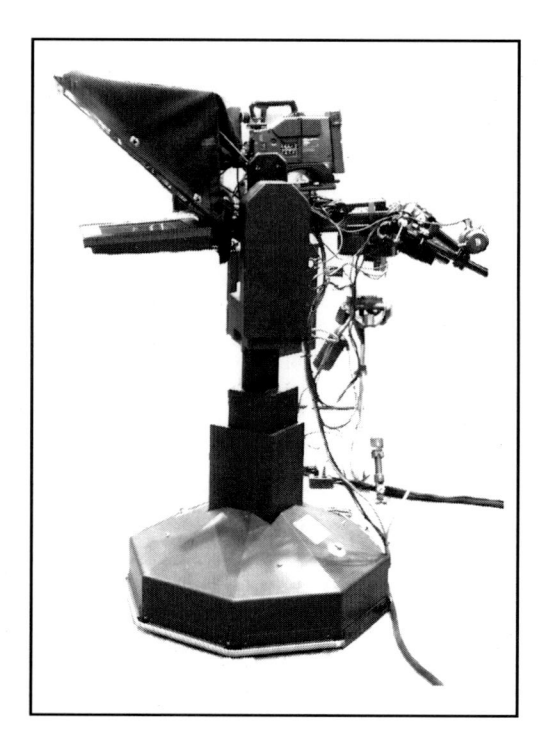

Steering is accomplished in one of two ways. In the parallel or "steer three" position, all three sets of wheels point in the same direction, and are turned in unison by a steering wheel connected via pins and rods to a chain connected to all the casters. In the tricycle or "steer one" position, only one wheel set is connected to the steering wheel. This allows you to twist or rotate the pedestal to a new orientation on the studio floor, permitting you to get closer to the edges of objects such as risers.

Robotic pedestals are also quite common, mostly used in automated newsroom environments where the shots do not vary from one newscast to another. These remote-controlled systems also include a robotic camera head that allows an operator to control tilt, pan, zoom, and focus from a centrally located control room panel.

Your Body

Many portable cameras for ENG/EFP use are designed to be carried on your body directly. The U-shaped section under the camera is theoretically shaped like your shoulder. Special body mounts or braces are sometimes used to distribute the weight more evenly over the shoulder and waist. The disadvantage of a body brace is that, unless the cameraperson is well practiced, breathing motion is transferred to the camera.

It's important to note that small consumer cameras (now often used because of their remarkable video quality) should almost never be used in a handheld way for professional broadcast quality results. The reason is that their light weight makes them too shaky for proper stability and framing. This is one of many things that distinguish professional-looking video from "home camcorder" or amateurish visuals.

If you must shoot handheld with either a full size broadcast camera or a consumer camcorder, do not zoom in! Zooming a handheld camera is guaranteed to produce shaking shots. Stay fully wide at all times. If you must get closer, and it is safe to do so, walk towards your subject.

Sandbags

For very low (on the ground) shots, sandbags allow a certain amount of camera motion, but enough stability that the camera can actually be left unattended, providing no movement is required. Sandbags also can be used for temporary attended stability when shots are required from less-than-solid positions, such as the tops of desks.

Cranes/Jibs

With a crane you can get from within half a metre of the studio floor to several metres above it. The camera can be swung left and right while panning and tilting, providing opportunity for some excellent effects and creative camera work.

The jib, on the other hand, can be run by one operator, and can do practically the same variety of moves as a full crane. Its controls also allow you to tilt, pan, focus, and zoom the camera. This takes a bit of practice, but jibs can be collapsed and moved almost anywhere, and they take little time to set up. They're very versatile, relatively inexpensive, and fun.

Steadicam®

The Steadicam® is a camera mount worn by the camera operator. Invented by Garrett Brown in the mid-1970s, it consists of a "vest" attached to an articulated arm system. Various springs in the arms absorb the wobbles and jitters while you run with the camera. In fact, when you run upstairs or on a mountain pass, your camera shots will come out as smooth as if you had used a camera crane. The Steadicam® system is quite heavy, and only experienced operators can wear it, along with its camera/monitor combination, for an extended period of time.

© Steadicam® photo courtesy of the Tiffen Company.

Helicopters and Flying Cameras

For overhead shots ranging over long distances, nothing can compare to a camera mounted in a helicopter. Whether it's a full-size chopper with a special camera and swivel mount in its belly (for example, Wescam™), or the completely radio-controlled and microwave-relayed video system shown in the illustration (Flying Cam™), it will be an expensive proposition. But sometimes, nothing else will do the job.

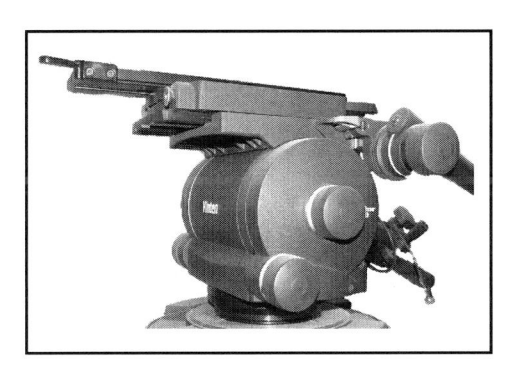

Camera Heads

A good camera mounting head should be able to counterbalance the camera easily and quickly (with its attendant accessories) by make the panning and tilting easy and smooth for the operator. All heads have panning and tilting locks as well as separate controls to regulate the degree of friction ("drag").

Usually confined to studio environments, the heavier-duty large camera head can counterbalance even the heaviest of studio cameras to permit smooth motion.

For lighter ENG-type cameras, fluid heads are often preferred. Fluid heads achieve the counterbalancing effect by having their moving parts operate in a thick oil. Their pan and tilt tensions can be adjusted by a selector ring.

Camera Operation and Aesthetics

Like riding a bicycle, operating a camera is something you learn by doing. There is no substitute for practice. Here we will discuss some operating principles and picture composition.

Checking It Out and Using It

How to Do a Camera Lineup

All cameras should be set up first thing each day and before each new location over the course of the day. With the modern technology available to us, this is commonly reduced to two basic steps:

- Put camera on lit chip chart; hit "auto white" and "auto black" switches.
- Shoot pictures. Have a nice day.

But could you do a camera balance manually?

Colour Balance

Set up the camera on a chip chart, so the arrows on the edge of the chart fit to the edge of the raster of the camera's scan. Be certain that the chart is lit evenly top to bottom and left to right. Use a light meter—or at the very least, zebra stripes—to do this. Set all colour correction controls to their preset or neutral position. Set the black level to 7.5 units and the iris so that the white chip just reaches 100 IRE.

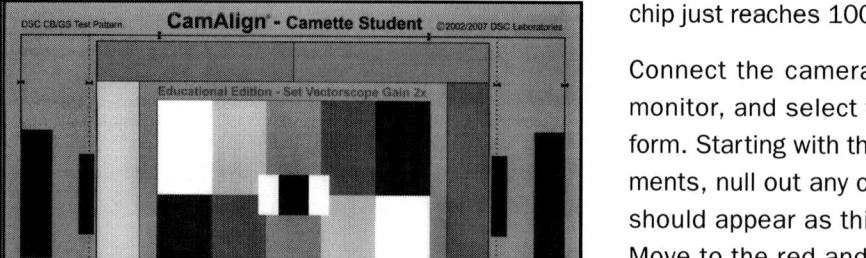

Connect the camera's video output to a waveform monitor, and select the FLAT response on the waveform. Starting with the red and blue black-level adjustments, null out any colour in the blacks—these chips should appear as thin as possible on the waveform. Move to the red and blue gain adjustments, and do the same. Go back to the black levels, and null again. Repeat until the chips have as little overall colour information as possible.

Why Bother Explaining This?

You may consider the above explanation to be redundant, given today's automatic lineup systems. However, for several reasons, it is a good idea to understand just what happens inside the camera when you press, say, Auto White.

Sometimes, even after doing automatic lineup, you will find that the shot just doesn't look "right." This is especially true in a multi-camera shooting environment when, even after automatically balancing a series of cameras, they don't quite match. The colour is not nearly what you expected, and so you will have to understand how to manually adjust the cameras. This is easily done by looking at each shot on a waveform monitor in the FLAT response position, slightly uncrushing the black levels, and nulling out the blacks and whites as necessary. In a situation like this, a qualified CCU, camera, or lighting operator will be required and expected to know all about camera principles and manual camera lineup procedures. It does no harm to be well versed in these methods.

Finally, today's HD video is of such high quality that directors and videographers are now creating film-like images. The understanding and use of camera controls to affect such elements as colour temperature, contrast ratio, image enhancement or softening, and much more is often required. This knowledge is now an implicit requirement for, and expectation of, the advanced professional cameraperson.

Studio Shoot

Before beginning a shoot in the studio, check your equipment:

- Is your intercom working (can you talk and hear)?
- Are your tilt, pan, drag, balance, locks, and counterweights (if using a pedestal) adjusted properly?
- Is your camera cable untwisted and unknotted? Can it uncoil easily?
- How's your viewfinder picture quality? Adjust it accordingly.
- How are your camera's optics? Does your lens zoom smoothly? Does it focus smoothly? Is your back focus correct?
- If you have a teleprompter, is it connected? Plugged in? Making a picture?

During the shoot, pay attention and think ahead:

- Preset your zoom so that when you land, zoomed in, you will be in focus. But do not zoom in and out needlessly during shots unless you are presetting your zoom lens.
- If you anticipate a dolly, ensure that your zoom lens is as wide as possible. This allows you to dolly with minimum vibration being transferred to the shot. The depth of field at this setting should be large enough that you should need to adjust your focus only when you land close to the object or event.
- When dollying and trucking, start slowly in order to overcome the inertia of the pedestal, and try to slow down just before the end of the move. If vertically oriented pedestal moves are to be executed, make sure that you slow down before it hits the stops at the extreme top or bottom of the pedestal travel.
- Be aware of stuff (and people) around you—other cameras, microphones, floor monitors, the floor director. A good floor director will help you to clear the way. He or she also will tap you on the shoulder to prevent you from backing into something.

- Watch for the tally light to go out before moving the camera, unless, of course, motion is intended. A good camera operator has the next shot lined up before the director calls for it. Your director will appreciate good visuals in an ad lib show, but do not try to out-direct the director from your position.

- Listen to what the director tells all the camera operators, not just you—you'll be able to coordinate your shots with those of the other operators.

- Avoid unnecessary chatter on the intercom.

When the shoot is over:

- Cap and lock the camera.

- Push it to a safe place within the studio.

- Coil the cable into a figure-eight loop.

EFP Shoot

Before you go, check your equipment:

- Are the batteries charged? Do you have spares? Don't forget to keep an eye on the microphone and mixer batteries.

- Does the camcorder actually make a recording? Have you done a test recording before you leave the equipment cage? Does the camera make properly colour-balanced pictures?

- Do you have enough tape or other recording medium? Does your recording stock have the little record protection punch-outs still intact, or slid in the right position? Are you sure you're using the correct type of tape for the camcorder, or the correct type of solid-state recording medium?

- Does your microphone work? Does your microphone cable work?

- Does the sungun work? Do you have a spare bulb? Do you have enough AC cords, bulbs, and adaptors, if you're using 110 volt lighting?

Consider the following as the seed of your own personal ENG/EFP shoot kit:

- extra recording medium (tape, solid-state)

- spare microphone

- small mic stand

- camera effects filters

- camera rain cover

- spare batteries for everything (camera, audio mixer, etc.)

- lights and stands, as required

- AC extension cords

- various clips, clothespins, gaffer tape
- small reflector for filling in dark areas of shots
- small white balance card
- lighting diffusion material and/or coloured gels
- small flashlight (comes in handy at times!)
- portable multi-tool (e.g., Leatherman, Gerber)
- pocket knife
- other tools as required

During the Shoot

- Set your equipment up. Turn on the camera. Power up the microphones if that's required. Do white and black balance. Repeat the balance each time you change locations where there is a significant lighting colour temperature change. If you are in environment with more than one light type, try to place the white balance card so it is lit by both sources.

- When shooting, try to keep the camera as steady as possible. Aim the camera with your whole upper body; have your legs absorb all the bumps. To reduce vibration, walk with the zoom lens in the wide-angle position.

- If you lose the action, don't worry: keep the camera steady, find out where the subject is, and aim the camera smoothly in the new direction. Alternatively, simply zoom out to reorient yourself to the new situation.

- When shooting under low light conditions, you will need to pay more attention to focus than when using a sungun or when in daylight because your depth of field is greatly reduced. Try to avoid fast movements. Your camera's inability to shoot reliably in low light will reveal itself as lag, smearing, or digital stepping in the video.

- Shoot with sound at all times. Even if you will be doing full post-production sound later and don't require the on-location sound, always record with the internal camera microphone at least. This supplies a good background sound source during the edit. This sound will also be invaluable in the post-production process for cues, takes, synchronization, and other purposes. Never come back from a shoot with a silent tape! When the talent is holding the external mic, don't run away from him or her to "get a better shot." Either you run together, or you must stay put.

- Above all, use common sense. Be aware of safety—yours and other people's. In ENG (electronic news gathering), reliability and consistency are more important than tales of adventure, no matter how spectacular.

After the Shoot

Unless the story is going immediately to air, take care of business first. Put away your equipment in the proper boxes or bags.

- Make sure your recording medium is removed from the camcorder!
- Ensure all switches are turned off. Don't forget microphone power switches.
- Cap the camera.
- Roll up all cables.
- Put your batteries in the charger.
- If the equipment got wet, confirm that everything is dried off thoroughly.
- If anything got broken or didn't work right in the first place, don't hide it from the equipment cage personnel. Get it fixed!

Filters

For many camera operators, a filter is just the little wheel on the side of the camera. However, with a little experimentation you can create a universe of film-like effects and images.

The basic video kit should include three groups of filters:

- correction, e.g., polarizing or graduated filters
- colour, e.g., sepia or pastel
- special effects, e.g., star or diffusion

Within these groups are a host of creative tools.

Most camera operators are familiar with correction filters, since the filter wheel on their camera has a neutral density, to cut bright sunlight, and a colour correction setting for daylight and tungsten lights. But there is no need to stop there.

For any high-contrast outdoor work, you should have a polarizing filter. It reduces the glare of snow banks and desert sand. It also cuts reflections from glass and metal, and makes transparent water shots in the Caribbean possible. In general, it is useful to darken a light blue sky and improve the contrast ratio between talent and background. The amount of correction can be adjusted simply by rotating the filter in the mount.

Colour correction filters range from those for fluorescent light, to UV reducing, to red and blue ones that combine correction and neutral density. Colour correction can be especially useful where the backlight and foreground light are of a different temperature.

A graduated filter, half colour/half clear, can be used both for correction and colour enhancing. A neutral grey will reduce the intensity of half the frame without adding colour. Darkening high-contrast areas such as sky or sand allows for better exposure of the primary subject. A blue filter brings out vivid colours in sky or water, and a pink or tobacco filter makes striking sunsets even in the afternoon. These are also helpful indoors to soften harsh overhead light.

Full colour filters can give normally "hard" video an emotional range and texture usually associated with film. Pastels evoke images from romantic firelight to nightmarish visions, blues can simulate mornings or evenings, coral filters can warm skin tones, orange can create forest fires. A didymium filter increases saturation of reds and oranges while softening blues and greens, by removing some yellow parts of the spectrum. One of the more widely used filters in video is probably sepia, a time machine that transports the viewer to the past.

Special effect filters can add a twinkle to an eye or generate special images. Star filters highlight points of light. Use them to accent candles, create sparkles on a client's product, or enhance an eye light reflected in the talent's pupil.

Travel to an altered state with a multi-image, prism, or vari-colour filter that changes colour as it is rotated. Roam from a misty morning to the depths of the bayous with fog filters. Journey to a world of dream or fantasy with diffusion filters. A centre-spot filter will make a client's product stand out in sharp relief by clearly positioning the object in a sea of colour or surrounded by haze.

Because of the increased sensitivity of current video cameras, filters can be used in combinations that multiply both the effectiveness and effects. For example, a pastel and a fog, or a diffusion and a multi-image can be combined.

With creative filtering, film makers have been emotionally manipulating audiences for over a century. These same techniques can be used on your next video commercial, story, or program.

10
LIGHTING

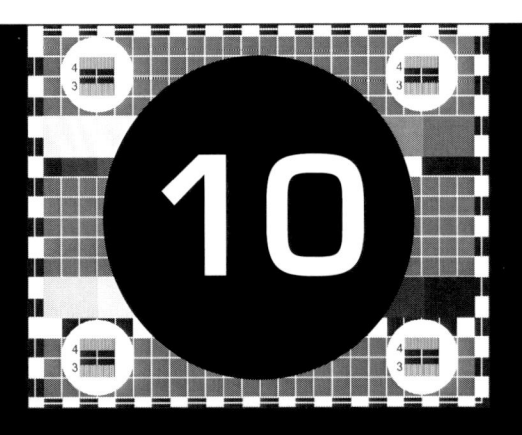

*"There are two ways of spreading light: to be
The candle or the mirror that reflects it."*

—Edith Wharton, "Vesalius in Zante"

Theatre Lighting: A Brief History

As television is the descendant of the live stage, it is worth our while to have a brief look at the history behind theatre lighting.

Gas Light

Philadelphia's Chestnut Street Theatre became in 1816 the first in the world to light the stage with gas. Gas provided brightness and control, but also unwanted heat, odour, and, unfortunately, fires. The successor to the basic gaslight fixture was the limelight. It featured a cylinder of lime (calcium oxide) that was heated to a high temperature by an oxygen/hydrogen gas flame. The limelight was the prototype of the modern spotlight. Invented in 1826 by Thomas Drummond, limelight produced a brilliant yet mellow white light, and was used in Covent Garden Theatre in London, England, beginning in 1837.

Electric Incandescent Light

During the 1840s many experiments aimed at the development of a workable electrical incandescent lamp were conducted. Nothing practical happened until 1879, when Thomas Edison developed a successful carbon filament incandescent lamp. Carbonized threads were tightly sealed inside a glass bulb from which all air had been removed by vacuum pumps. A voltage was then applied to the ends of the filament, and steadily increased until the white heat (incandescence) reached a stable, bright glow. The first incandescent electrical lamp lasted 40 hours, at which point the question was first asked, "How many inventors does it take to change a light bulb?"

That invention marked the birth of both electric lighting and the electric age. Edison's obvious next step was to invent electric generators to supply his lamps with current. In 1882 his Pearl Street generating plant began serving almost sixty customers in the Wall Street district of Manhattan by supplying electric current for more than a thousand lamps.

On a stage, electricity—specifically an electric-arc-light—was first used at the Paris Hippodrome in 1878. While this was, strictly speaking, an electrical fixture, it created light by generating an electric spark between two carbon rods. London's Savoy Theatre in 1881 became the first to use incandescent filament lamps. Electric filament light allowed more variety and control, and was the perfect complement to naturalistic staging, which developed at the same time.

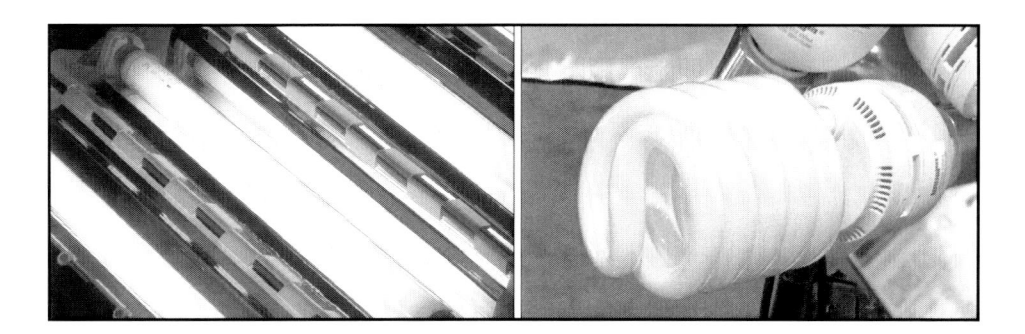

Electric Fluorescent Lighting

Fluorescent light, first exhibited at the Chicago Centennial Exposition in 1933, came into popular use around 1939 and by the 1950s had largely replaced incandescent lighting in schools, offices, hospitals, factories, and commercial establishments.

Fluorescent light belongs to the group of lighting devices known collectively as discharge tubes—glass tubes filled with metal vapour, with electrodes at each end. Electric current eventually ionizes the vapour, which begins to glow, producing light. Neon light is the most common use of this type of lighting. Discharge tubes are also widely used for street lighting. For example, high pressure sodium vapour lamps produce a brilliant yellowish orange glow that is used to light major highways throughout the world. Mercury vapour lamps, which operate at higher pressures, produce a greenish blue light.

A fluorescent light is a highly efficient light source that uses a mercury arc and a fluorescent phosphor coating inside its tube. The mercury arc in a fluorescent lamp operates between two heated coils called cathodes. Much of the arc's energy is emitted in the invisible ultraviolet range, but the phosphor coating in the tube converts it to visible wavelengths. Different phosphors can produce a range of colours varying from cool to warm.

The advantages of a fluorescent light are its high light output per watt and its long life. A 40-watt fluorescent lamp produces more than 70 lumens per watt, and lasts 15,000 to 20,000 hours, depending on the burning cycle. By contrast, an average tungsten studio bulb lasts only about 250 hours and produces 25 lumens per watt, and a domestic 40-watt incandescent lamp produces only 12 lumens per watt and has a rated average life of 1,500 hours.

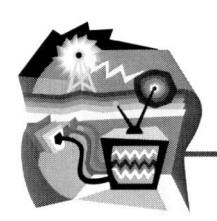

LED (LIGHT EMITTING DIODE) PRINCIPLES

LEDs are what is known as a semiconductor technology. This means that they are made up of two layers within the LED itself, called N-type and P-type material. Each one of these layers is attached to an electrical connection. A common material used in LEDs is aluminum gallium arsenide (AlGaAs).

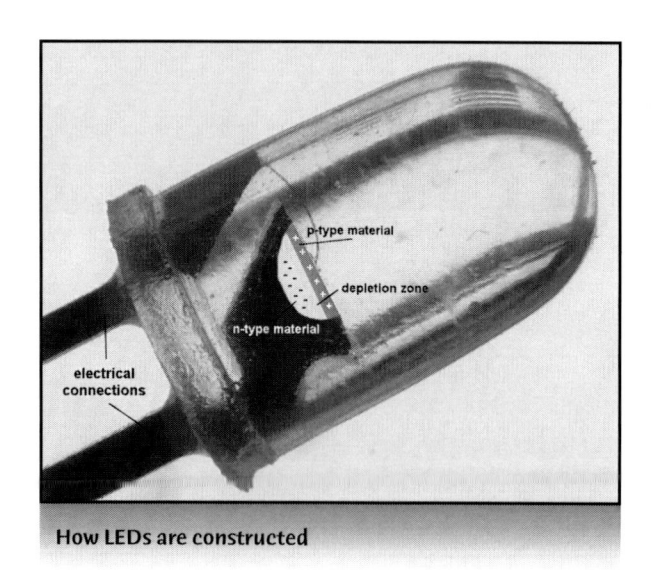

Because the two layers of material are "doped" (i.e., contaminated with impurities), the N-type layer has excess electrons within it, and the P-type has what are called "holes." Since materials tend to want to move towards a state of electrical equilibrium, the semiconductor creates an area of neutrality between the N-type and P-type layers known as the "depletion zone." (Semiconductor theory observes that the holes are, interestingly enough, also attracted towards the electrons!)

Direct current (DC) is applied to the LED to create light—as in a flashlight, for example. The positive terminal of the source of electricity is connected to the P-type material, and the negative terminal is attached to the N-type side of the LED. This forces electrons from the N-type material, and holes from the P-type material, to be pushed towards one another, filling in the depletion zone. During this process, the electrons release energy in the form of photons—light.

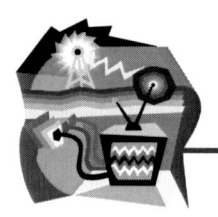

How LEDs are constructed

MEASURING LIGHT

Footcandles, Lumens, and Lux

A footcandle is roughly the amount of illumination on a surface one foot away from a common candle. The metric measurement of light is lumens per square meter, or lux. While this measurement is, strictly speaking, a two-dimensional one (the light measured across a one meter by one meter surface), one lux, as a term used in the television and film industries, is more commonly thought of as the amount of light falling on a surface one metre away from a typical candle. Measurement of light in lux is now more common, although the footcandle measurement system is still used in some areas of North America, particularly the United States. If you wish to convert footcandles to lux, or vice versa, the ratio is approximately 10:1—ten lux is the equivalent of about one footcandle.

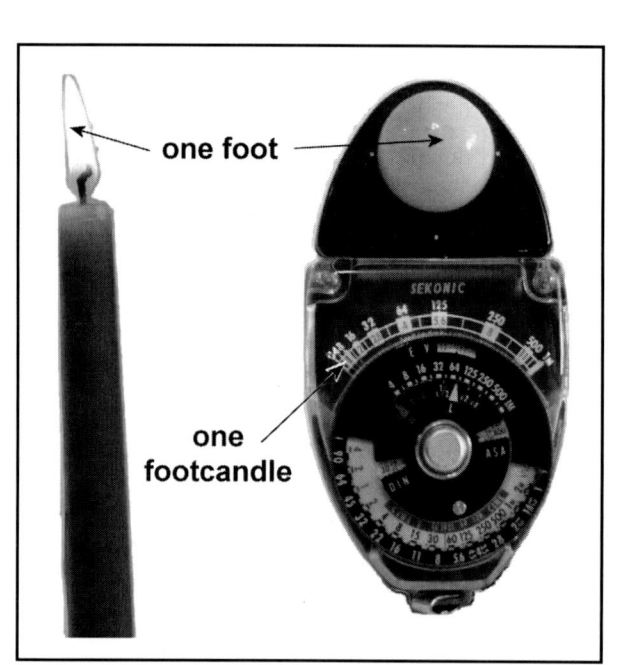

Light Meters

We measure light in studios with a light meter or exposure meter. The original Weston light meter used a selenium photoelectric cell that produced an electric current proportional to the lighting reaching it. This, in turn, was measured directly by a sensitive electric meter movement. Another type of analog light meter employs a cadmium sulphide photoconductive cell, which changes electrical resistance with light level changes. This varies the amount of current from a battery inside the meter, and this varying current is then sent to the meter movement. Digital light meters are now common and can read light levels in either lux or footcandles.

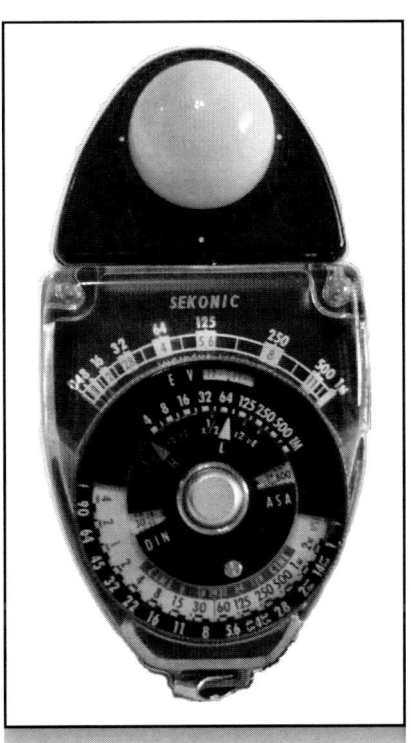

Analog photometer

Digital photometer

Inverse Square Law

Perhaps you've heard of the inverse square law. If you have a light source that radiates uniformly in all directions (a candle or a regular home incandescent bulb, for example), the light intensity will fall off as the "inverse square of the distance." An example will help.

If the intensity of a typical candle is 1 lux measured at a distance of 1 metre away from the flame, its intensity at 2 metres will be 1/4 lux (1/2 × 2). At 3 metres, the intensity will be 1/9 (or 1/3 × 3) of a lux. And so on.

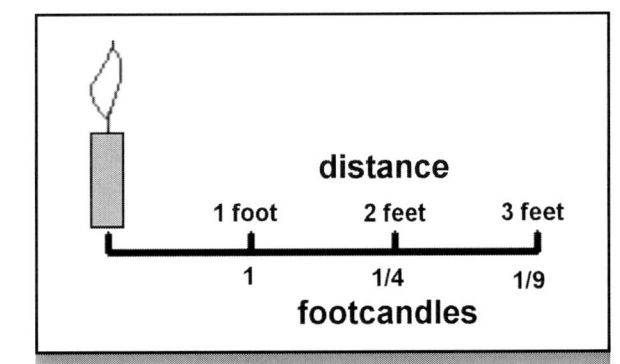

The inverse square law: light intensity falls off as the 'inverse square' of the distance

However, this principle doesn't hold for fixtures like Fresnel spotlights, flashlights, laser beams, or car headlights. These are all focused fixtures; therefore the drop-off is much less noticeable. This feature is known as the columnation of the light source (i.e., how much the light is like a column). The best columnated source is a laser beam, which maintains its intensity over great distances.

LIGHTING FIXTURES

Spotlights

Fresnel

The most popular spotlight is still the Fresnel, so named after Augustin Fresnel, who invented the tiered lens that it uses. The Fresnel spot is quite flexible, providing high light output and the ability to focus the beam in a flooded or spotted beam. Focusing can be done in a couple of different ways. One involves moving the bulb and back reflector unit by a turnscrew or quick sweep focus handle. The other is by moving the front lens instead of the hot lamp element—this is called a ring-focus spotlight. Fresnels come in a range of sizes from 100 watts up to 20 kW (kilowatts).

Fresnel spotlights

Interior of a Fresnel fixture

Ellipsoidal

This is the hard-edged spotlight that creates clearly defined pools or circles of light. These circles can be modified by adjusting built-in shutters to create rectangles or triangles. Some ellipsoidal lights also can be used as pattern projectors by inserting a cookie (short for cucalorus)—a metal pattern cutout or glass plate) next to the shutters. These patterns can be placed in sharp or soft focus, creating a wide variety of effects.

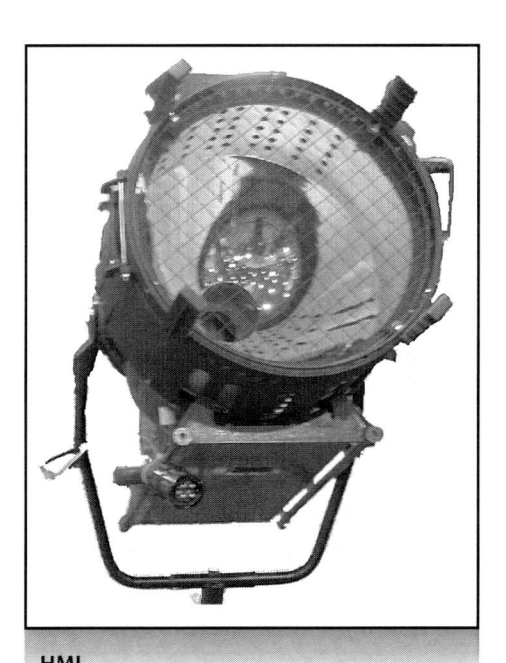

Ellipsoidal spotlight

Cookies

HMI

The Halogen-Metal-Iodide light is an arc-lamp Fresnel fixture. It delivers from three to five times the illumination of a quartz instrument of the same wattage, uses less power, and develops less heat. They come in a wide range of sizes starting at 200 watts. They are more expensive however, and each fixture needs a special starter-ballast unit to get it going. These boxes get warm and occasionally hum. They're unsuitable for fast newsgathering work or short turnaround situations, since they take a minute or two to build up to full illumination.

HMI

Floodlights

Scoop

The scoop has no lens, just a large scoop-like reflector. The simplest, fixed-focus, version has no controls at all, short of dimming the source or placing scrim material in front of it. While these are handy for filling in shadows in tighter areas, they have been largely replaced by broadlights and softlights.

Broads and Softlights

These are versatile fixtures, providing very even lighting. Softlights have large, white reflecting surfaces to provide extremely diffused, even lighting. They are useful if you want to increase the basic light level without affecting your carefully controlled highlights and shadow areas.

Broads act like a series of scoops—they evenly light a large area with somewhat diffused light. Some broadlights have barn doors; others may have focusable beams. You can attach an "egg crate" to the front of a broad, a multi-celled director that minimizes the amount of spill created by the broad.

Broadlight

Floodlight Banks, Strip or Cyc Lights

Floodlight banks are high-intensity reflector bulbs mounted in a six, nine, or twelve spot arrangement. They are used mostly on remotes or outdoors, to illuminate large areas over a long distance or to act as a booster to reduce harsh daylight shadows.

Strip, or cyc, lights are used to achieve even illumination of large set areas, such as the cyclorama or other uninterrupted background area. You can use them on the studio floor or suspend them in the grid. They are also used for silhouette lighting, where the background must be evenly illuminated, and the foreground must remain unlit. Today, many new cyc light strips are being manufactured using LED technology.

LED cyclorama lights

Fluorescent Lighting Fixtures

There was a time when fluorescent lamps lacked the flexibility of incandescent lamps in that the wattage couldn't be changed and the lamps couldn't be dimmed. That's changed dramatically over the past few years with a number of manufacturers like Kino-Flo and Videssence making products that, while still soft in character, allow on-air talent (in news programs, for example) to keep their cool while under the no-longer hot lights of the set. New fluorescent tubes are being manufactured as "tri-phosphor," meaning they have red, green, and blue phosphor compounds that glow to produce light similar in colour rendering to either incandescent illumination or daylight. The final drawback of fluorescent light, that it flickered noticeably on camera, has also been overcome by use of high-speed switching power supplies. The lights still flicker, but the pulsing rate is much higher than television or film frame rates, and therefore it is not visible on camera. As well, fluorescents can now be dimmed using these same high-speed switching systems.

Strictly speaking, they're a soft light, but at close range they can be used as a harder type of illumination.

LED (Light Emitting Diode) Lights

Lighting fixtures with high-intensity LEDs are quickly growing in popularity. Not only do LEDs have low power consumption, making them ideal for EFP work, but different coloured LEDs can be packed into one lighting unit. This ability to create any colour means that a lighting designer no longer has to have a large number of coloured gels on hand. Because LEDs run on direct current (DC), they can be easily dimmed. They generate a fraction of the heat of regular incandescent lights. The fixtures can be manufactured into original shapes and sizes, so that they can be used in many different applications. As technology advances, these lights will become increasingly common in television productions.

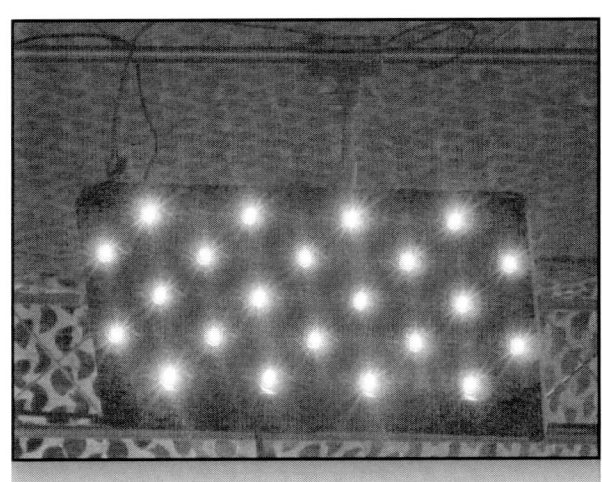

LED grid light

LED camera lens ring light

Mounting Lighting Equipment

Studio lights are usually hung from either fixed pipe grids or counterweighted pipe battens that can be lowered and raised to convenient positions. They are fixed to these pipes by C-clamps or pantographs (spring-counterbalanced scissor-like devices).

Lights also can be mounted on floor stands, portable or otherwise. These can support any type of lighting instrument and can be vertically adjusted to a convenient height.

C clamp mounting

Floor stand mount

LIGHTING CONTROL

Barn Doors

This simple but effective control is useful if you want to block certain areas from illumination. They are also used to prevent the back light, used on sets and talent, from shining into the camera lens. Because barn doors slide into their fixtures easily, they also slip out of them just as easily. Therefore, it is important to ensure that they are safety chained to the light fixture itself. When adjusting them, wear gloves since the black surface of a barn door gets very hot.

Barn doors

Flags

These are rectangular frames with stretched cloth, or sheets made of metal or wood in various sizes. They can be put anywhere on the set where you want to block light, and are mounted on flexible arms to allow complete freedom of adjustment.

Flags

Reflectors

Highly reflective sheets of metal, tin foil, white cardboard, or painted wood can be used to bounce light back onto the scene. They can be used as a substitute for fill lights on remote or outdoor locations. Experimentation reveals their full potential.

Various reflectors

Scrims

Scrims are spun-glass diffusion material that you place in front of floods or spots. They absorb a lot of light, but provide quick diffusion in a variety of situations. Scrim material comes in many densities.

Scrim material stretched over a frame

Eggcrates

Eggcrates consist of several (or sometimes dozens) of rectangular boxes built into one unit, which make the illumination from softlights directional. Softlights would otherwise tend to throw their light in many directions.

Eggcrate over a broadlight

Softboxes

Softboxes are large fabric boxes or bags that attach onto the front of hard or soft light fixtures. Often softboxes are so large that the fixtures (or bare bulbs) fit entirely inside them. They allow an otherwise hard light fixture to be extremely soft and even.

Various softbox types

Dimmers

The most flexible intensity control is the dimmer. By allowing more or less current to flow to the lamp, it illuminates the scene with a higher or lower level of light. Dimmers can be used for intensity control, quick illumination changes, colour changes (if various fixtures are equipped with coloured gels), and special effects lighting such as silhouettes, pools of light, and day-to-night changes.

Individual dimmer controls can be arranged so they have independent control over a fixture, or they can be grouped under the control of a submaster or master, so that several lighting fixtures can be dimmed together at once.

Lighting dimmer control panel

Lighting control systems also include either a hardware or software (computer controlled) patch panel, allowing various fixtures to be controlled by different dimmers, at the whim of the lighting operator.

Television Studio Lighting

Why do we light people and objects with special fixtures in television? Obviously, we have to provide proper technical exposure for the camera, so we can get clear pictures with good contrast and depth of field. Aesthetically, we want to make the people and scenery appealing; indeed, sometimes we want to make a set look like more than it is, or to set a mood.

In general, furnishing good illumination involves far more than just providing a certain number of lux on a scene. Light can model objects or flatten them, reveal colours or distort them, provide a cheerful environment or a gloomy one. Reducing glare can increase visibility. In addition to calculating illumination, the lighting person must deal with these problems through the choice of a particular type of light source and the placement of the fixture.

BASIC TELEVISION LIGHTING TERMS

When lighting people, we frequently refer to the key, fill, and back lights.

The key light is the apparent principal source of directional illumination falling on our subject. It is generally adjusted to have the highest intensity and may be a "hard" or "soft" type of fixture.

Low key lighting

High key lighting

The fill light is supplementary illumination used to reduce the shadows or contrast range, especially those produced by the key light. It, too, can be a "hard" or "soft" fixture.

The back light is for the illumination of the person from behind, to produce a highlight along the outline of the individual; this sets the subject apart from the background. To create this effect, this light is usually a "hard" type of fixture.

High-key and low-key are two terms used to describe an overall lighting "look." High-key lighting has predominantly high-intensity areas, with a background illumination that is similar in intensity to the subject. Low-key, on the other hand, has mostly lower light levels, with backgrounds even lower. The subject illumination may even be restricted to only certain areas of the individual. The lighting on a game show is often high-key; most soap operas are low-key.

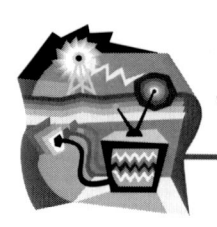

TELEVISION'S LIMITS

Contrast, Two-Dimensionality, Colour

Contrast

Below a certain light level, the television camera gives us back no pictures. Above a certain level, we also have no pictures, because everything in the scene is too washed out. Between these extremes we have a ratio. Some engineers and camera manufacturers say it's 30:1; others mention numbers as high as 50:1. Motion picture film, by comparison, is often considered to have a contrast ratio of up to 400:1.

In our contrast ratio calculation, we also haven't accounted for home viewers placing their brightness and contrast controls in all sorts of settings. So, let's stick with 30:1 to allow for variations within the television system and inside the home receiver. Practically speaking, what does this mean? As an example, if you have peak light levels on a scene reading 1500 lux, nothing you want visible in the same scene should be below 50 lux, within the same shot. This doesn't sound like a great limitation, but we are talking about the extremes of the 30:1 ratio here. For best results, your visible scene information should not approach these limits.

As well, you should be aware of the general reflectance of the objects you're shooting. Highly reflective objects (glossy, light-coloured set furnishings, for example) need less light than matte, light-absorbing surfaces (e.g., velvet drapes in the same set).

Try to avoid extreme brightness contrasts in the same shot. If you are shooting a highly reflective object (for example, a diamond ring), place it on a medium light cloth, not velvet. This way you can limit the amount of light hitting your key object without making the cloth appear too dark and muddy.

Lighten shadows with fill lights. This will show off detail that otherwise would be hidden and will also reduce the overall contrast on the set.

Television's limited contrast ratio is the reason that talent shouldn't wear a lot of pure white or black. A crisp white shirt or white mohair sweater (or, conversely, "black tie," or "goth" black attire) looks great on the street. However, it can be difficult to get skin tones right and simultaneously show off these clothes well when viewed on television.

Depth

Because the television screen has only two dimensions, the third dimension, depth, must be created by illusion. This two-dimensional problem is why chroma keys so often look like paper cutouts laid over top of a background; there are very few, or sometimes no, clues about the depth of the objects or persons being keyed onto the background image.

A proper control of light and shadow is essential for revealing the actual shape and form of three-dimensional objects, their position in space, and their relationship to one another. It is often the shadows, rather than the light, that show the form and dimension of an object. Lighting that emphasizes shadows can make a relatively smooth surface look richly textured; conversely, soft lighting can smooth rough surfaces.

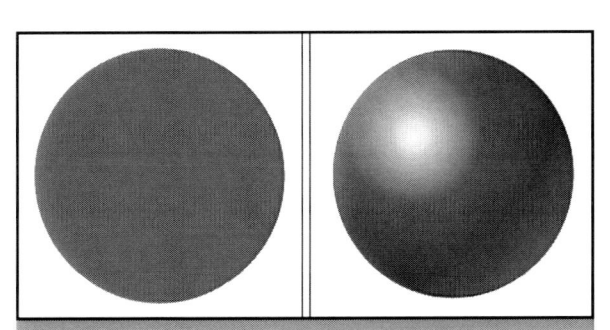

Two images of a ball, with the second image using effective lighting to illustrate three dimensionality

The lighting of the set also must provide the illusion of depth by outlining performers and objects against the background. Back lights work best; high contrast ranges between the background and these objects or people are also effective.

Colour Temperature

Colour temperature is the standard by which we measure the relative reddishness or bluishness of so-called "white" light. This should not be confused with the brightness, or intensity, of the light. Candles give off warm light; daylight gives off a "colder" light. These differences can be measured in degrees Kelvin.

The standard colour temperature for television lighting is 3200 degrees Kelvin—fairly white with a slight warm tinge. Most studio lights are rated for this colouration. Some lighting fixtures designed to augment outdoor lighting, however, are balanced for 5600 degrees.

When you dim a lighting fixture, either in the studio or at home, your eyes quickly adjust to its lower colour temperature. Colour cameras are not so forgiving; they don't adjust at all. Applying between 25% to 100% power to an incandescent fixture will change the colour temperature ±150 degrees Kelvin. As a result of this phenomenon the argument persists that you should never dim a studio fixture. As it turns out, you can, in fact, dim a fixture by up to 10% without having the colour change become too noticeable. Not only does this reduce power consumption, but it almost doubles the life of your incandescent light bulbs in those fixtures.

One interesting note about the new fluorescent fixtures and their dimming properties is that they do not change their colour temperature as they dim, because of the system used to reduce the light intensity of the tubes. The same is also true of LED fixtures.

Notice particularly the vast changes within the first two hours after sunrise, and during sunset. Early morning and afternoon/evening shooters, take note—your colour temperature will change frequently during these times of the day, so be ready to frequently adjust your camera's colour balance.

If adverse lighting conditions are all you have to work with, make sure your white balance is accurate—do it as often as you feel necessary. In addition to this, you may need to use one of the built-in filters on the camera filter wheel. If you move from bright outdoors to tungsten indoors, change the wheel back to an indoor setting. Some camera people do this on the fly, as they walk from one environment's colour temperature to the other. In "live" situations, this is entirely acceptable.

	Degrees Kelvin
sunrise	1000
candlelight	2000
incandescent light, and also one hour after sunrise	2500-3400
overcast sky, mid-morning and late afternoon sun	6500
hazy sky	9000
bright sunlight at noon, no clouds	15000

Typical colour temperatures

Gels

If you are in a shooting environment where there are many windows, orange gel material can be attached to the window frames to convert daylight colour temperature to tungsten. It comes in various ND (Neutral Density) ratings as well to reduce the light intensity streaming in from outdoors. If there are too many windows to do this, or if they are too large, you can colour correct your tungsten lighting fixtures to daylight using colour correction blue gel. However, this does cut down on the amount of light your fixture will now be able to throw on your scene.

Gel material, of course, doesn't just come in colour correction or neutral density selections. An entire range of colour is there for you to use in creative ways.

Lighting fixture with colour correction gel

Reality/Non-reality

We can create illusions in our television scene by thinking about what reality looks like. Long shadows suggest late afternoon or early morning; bright light establishes "outdoors" environments. A periodic flashing light through venetian blinds conjures in our minds a seedy hotel or neighbourhood.

Likewise, we can create moods using similar techniques. Reflections of light dancing off of water can create calming effects. And of course, everyone's seen the "lighting from below" look of B-grade mystery films. Watch television shows and films and take note of how they achieve specific moods and genres with light.

In a word—experiment!

11

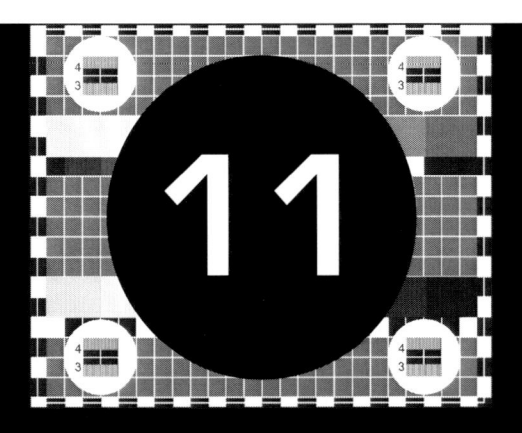

SPECIAL EFFECTS

"You know, television was actually invented in the 1890s, but they couldn't get it to work until the '40s when they came out with gaffer tape."

—Walter Pyle, CBS

Switchers

Up to this point, we have discussed some of the individual sources available as production elements. Live television production involves more than one of these sources, and so a method is needed to transition from one video source to another, or "join" them together. This device is a switcher, and it is the video equivalent of the audio console.

Switchers can be simple or extremely complex in their production function and electronic design. All switchers, however, perform the same basic functions: selection of the proper video source from several inputs, basic transitions between two or more sources, and the creation of or access to special effects. Some switchers, called audio-follow switchers, also take the program audio associated with a video source; these are used in master control situations and will be discussed later. As an aside, it is important to realize that the video sources themselves do not actually go through the switcher panel that the operator uses. Instead, the panel can be thought of as a remote control that directs the actual switching electronics housed in a separate rack-mounted unit.

Analog Switchers

Cutting

Switchers have rows of buttons called buses, and each video input on a switcher has its corresponding button. Cutting from one source to another involves simply pressing the button assigned to the source. The switchover happens only during the vertical interval of the two synchronous video signals, so there is no perceptible jump in the picture when cutting. If all you wanted to do is cut from one source to another, a single bus would do. We usually require more sophisticated effects, so a slightly more complex switcher is required.

Dissolving

To create a dissolve, we need to use two different sources at once. This will require two buses, and some sort of way to transition between them. A fader bar is the conventional way of controlling the dissolve rate. To dissolve, we select the two different sources on the two buses, and push or pull the lever to position it towards the other bus. When the lever is in the middle position, the output is in a "partial dissolve" or "super." We can still do our cuts with our two-bus system, by pushing any of the buttons on the bus that is live to air.

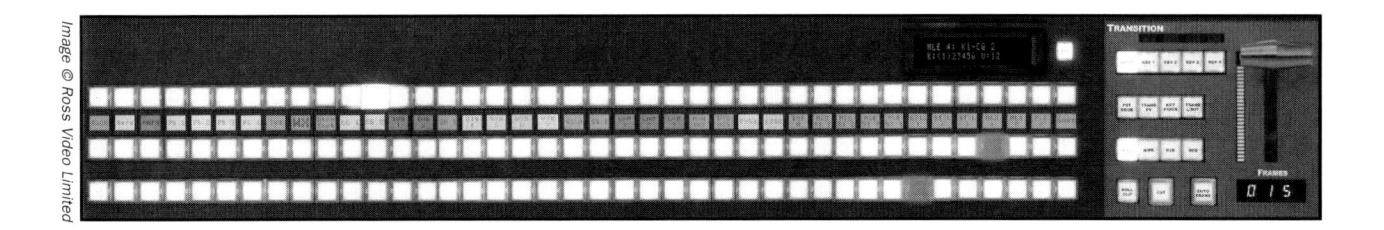

Dissolving involves sending control voltages from the switcher panel (which vary depending on the position of the dissolve lever) to the remote switching electronics. Two voltage-controlled amplifiers within that equipment change their output level up or down to correspond to the control voltages. This creates the dissolve, which is really one source increasing in level while the other source is decreasing. This is the equivalent of a cross-fade in audio.

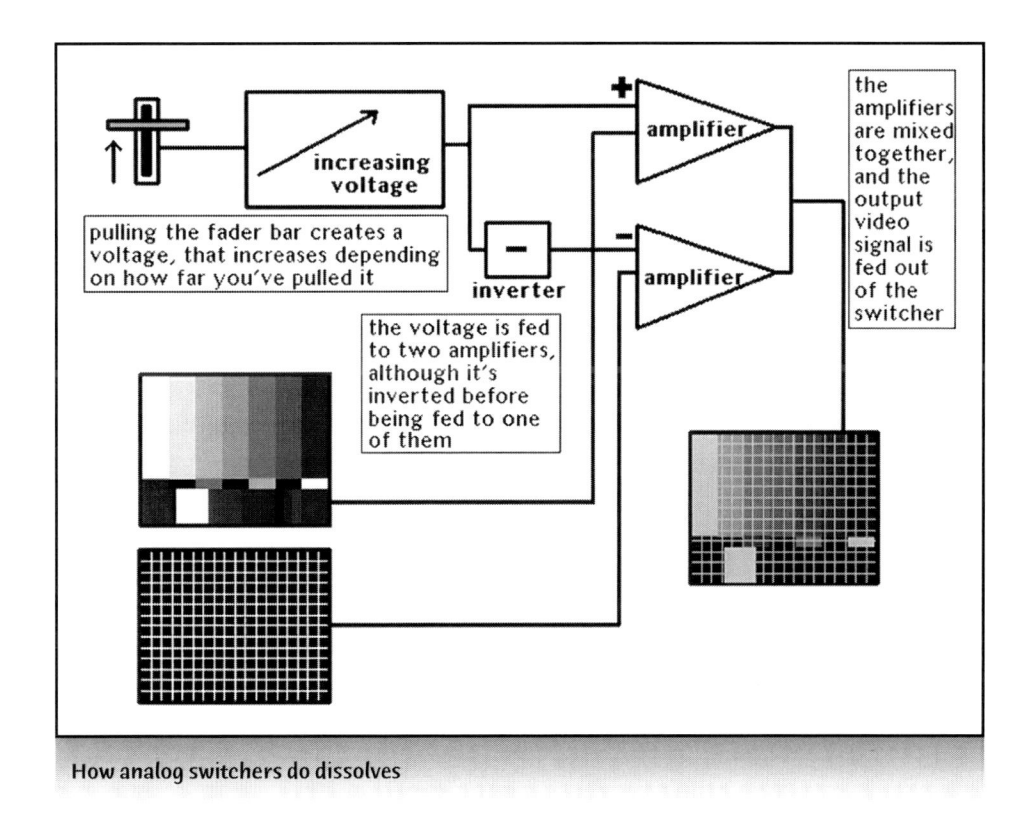

How analog switchers do dissolves

Wipes

Wipes are a more sophisticated transition, when one television picture is removed with a relatively hard edge and another is revealed. They can be done using the same two buses. To add this new feature, there are two buttons labelled WIPE and DISSOLVE, so our fader bar can do double duty. Our bus pair is now called a mix/effect unit, M/E, or MLE.

How analog switchers do wipes

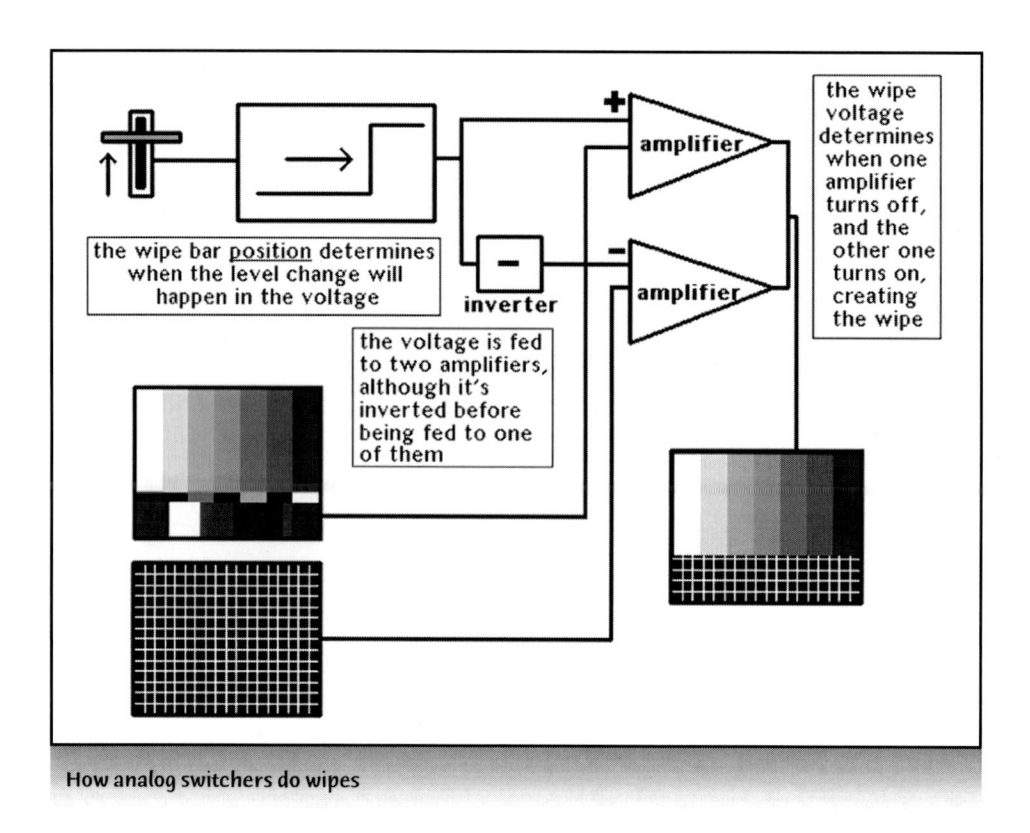

To understand how wipes are done in an analog switcher, consider a simple one, a wipe from the top of the screen to the bottom (a horizontal bar wipe). When the fader bar is at the topmost position, the entire screen is taken up with the top bus's source. As the bar is moved down, the switcher detects the voltage change sent to it from the fader bar. Based on the new levels, it decides at some point during the field of video being displayed to cut instantly over to the second source. So, the voltage level from the fader bar is used for positional information on the screen. This cutover point is what is seen as the edge of the wipe transition. As the fader bar moves more, this cutover point moves farther down the screen. Eventually, when the bar is at the bottom of its travel, the entire picture has been replaced by the second source. A simple vertical bar wipe is similar to the horizontal one, except this time the voltages are interpreted as positional information for a series of switchover points, one on each horizontal line. The further the fader bar moves, the more the second source is revealed on the screen.

Patterns

Combining the actions of vertical and horizontal wipes gives you the ability to do corner wipes—those involving both left-to-right and top-to-bottom motions at once. These wipes, and usually many more variations, can be individually selected by a matrix of buttons on the switcher called the pattern selector. Each button usually has a small picture on it representing the type of wipe that's selected.

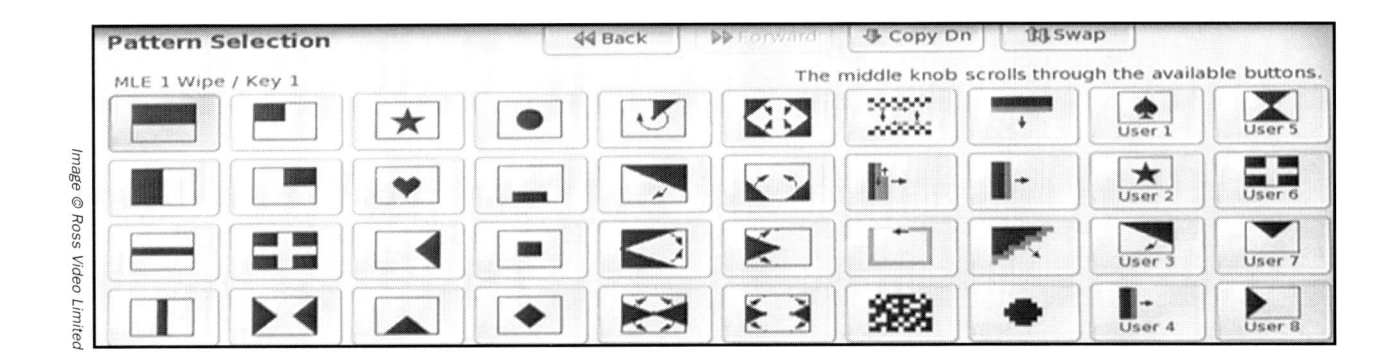

Pattern Modifications

The wipe transition doesn't always have to be a hard edged one. Consider that in a dissolve, no voltage present from the fader bar means only the first picture will be present, and full voltage means that only the second video source will be shown. In a wipe, the voltage is converted to a position on the screen. The ramped effect of a dissolve can be combined with the positional information used in wipes to create a wipe edge that quickly dissolves from the first image to the second. This results in a soft-edged wipe.

The wipe doesn't always have to have a straight edge, either. You can further influence the wipe pattern by introducing to it an additional signal waveform. This creates modulated wipes with soft, curving lines (a sine wave generator), teeth-like lines (a triangle or sawtooth wave generator), or squared-off lines (a square wave generator).

Wipe Direction

There's no rule that the top of the fader always represents the top of the video picture, or the left of the picture. Sometimes you'll want to do a wipe in the reverse direction. To do this, an additional direction button is made available, usually labelled as FORWARD/REVERSE. This just reverses the polarity of the voltages sent out from the fader bar. A third button is sometimes available on the switcher labelled FLIP FLOP, which sets up the switcher to change automatically between Forward and Reverse modes with every wipe transition. Also, many switchers have a joystick positioner, which allows you to vary on screen the starting and ending positions of the various wipes.

Auto Transition Buttons

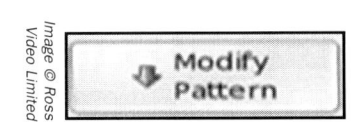

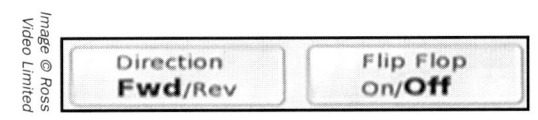

We often don't actuate the transitions with the fader bar. Switchers come with two additional buttons: AUTO TRANS and CUT. The AUTO TRANS allows us to do smooth dissolves or wipes to a predetermined timeline (measured in duration of video frames). The CUT button causes the source on the preset bus to cut to air. As the new source is taken to air, the previous one is automatically positioned on the preview or preset bus, thereby showing up on the preview monitor. It's then ready to be taken to air at the next press of the CUT or AUTO TRANS button. This sort of transition type makes a two- or three-camera interview cutting situation very easy, with minimum hand reach over the live buses.

Split Screens

If you stop a wipe halfway through its transition, it becomes a split screen. A special PRESET PATTERN button is available that allows you to vary the size of various wipes using an auxiliary set of controls instead of the fader bar. These controls, when used with the joystick, allow you to set up commonly used stationary boxes, circles, angles, corners, and other effects without tying up a fader bar in mid-position. With the use of a SYMMETRY control, the aspect ratio (width to height ratio) of certain split screen wipes can be changed. For example, a circle wipe also can be used as an ellipse, or a square box wipe can be elongated horizontally or vertically.

A special type of split screen effect is called a Spotlight. This effect looks like a circle wipe (with either a hard or soft edge), except that it lets the base picture show through, albeit darker than the highlighted area. It is used to draw attention to a specific portion of the screen, as though you were shining a spotlight on it.

Keys

Keying refers to electronically cutting out portions of a picture and filling them in with another image. The basic purpose of a key is to add titles to a background picture, or to cut another picture into the background. The cutting process is electronically similar to that used for wipes and split screens. However, instead of an internally generated control voltage, information from an external video source determines when to "cut over" from one source to another.

As video pictures do not usually contain extremely hard edges, an extra process is used to create the hard edge used as the transition marker. The video that will be used as the pattern for the hole goes through a processor that compares its luminance level at any particular moment with a reference voltage. If the luminance in the video signal is lower than the reference, the background picture remains on the switcher output. If the luminance in the keying video is greater than the reference, the second picture will be allowed to cover the background picture. This reference voltage can be varied using controls labelled KEY GAIN and CLIP. The Key Gain allows more or less of the keyed video to show through onto the background image depending on the luminance value at any one point in the keying video. The Key Clip control adjusts the "hardness" of the keyed edge between the two sources.

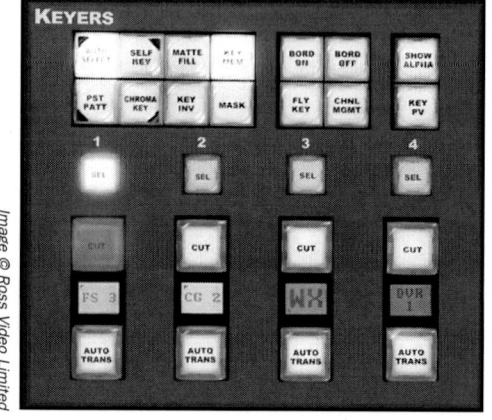

How analog switchers do keys

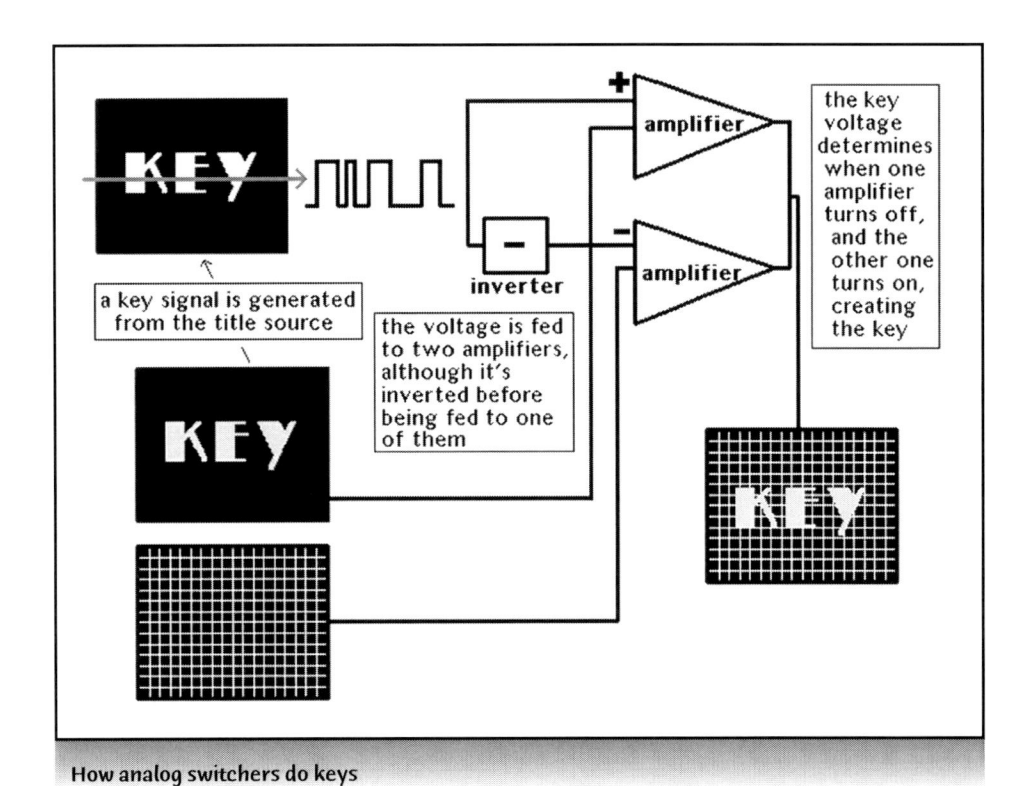

There are various types of keys. If the cutout portion of the base picture is filled with the signal that is doing the cutting, it is called an internal key, normal key, self key, or luminance key. If the cutout portion is filled by an external source, it is an external key. Sometimes the hole is filled with an electronically generated colour from the switcher; this is a matte key.

Normally, the keying action of a switcher is such that the lightest parts of the keying video create the hole to be filled. If the reverse situation is desired, most switchers come with a KEY INVERT button, which reverses the polarity of the key. For example, with a normal white graphic keyed over a background, the white graphic appears over the background video, and the black graphic background is eliminated. Depressing the KEY INVERT button provides a "see through the graphic" effect. That is, a piece of the previous background scene in the shape of the white graphic letters is keyed over the graphic's black background.

Normal and inverted key types

Sometimes, keying video has noise or other artifacts around the edges, marring the visual effect desired. Some switchers have an independent mask pattern generator that limits the display area of the keying signal. The boundaries of this pattern generator are often in the shape of a simple rectangle, and are manipulated by four rotary controls that vary the size, shape, and position of the masked area.

Chroma key is a special effect that uses colour, instead of luminance, for keying. A control on the switcher selects which colour will become transparent during the keying process, thus allowing the background video to show through. Usually green is selected if human performers will be in the shot. Historically, blue was used, since it was most opposite in chroma phase to skin colour. The proliferation of blue jeans in the 1960s, and the fact that occasionally blue-eyed talent would have their eyes disappear, gave way to using a bright green as the most common hue. Chroma key can, however, be used with any colour you wish.

Chroma keys come in two forms: RGB and encoded. In encoded chroma keys, the colour selection process picks a particular phase of subcarrier to eliminate. However, since video pictures seldom have pure phase colours in them, and chroma resolution is usually reduced by the NTSC encoding process, the results are sometimes less than perfect.

| background | foreground (camera) | chromakey composite |

The chroma key process

In RGB chroma keys, the separate red, green, and blue channels from a source (often a camera, although other sources with these outputs can be used) are brought into the switcher directly. From these separate channels the colour to be keyed out is selected. This has the advantage of cleaner keys since, to select a particular colour for keying, only a fairly straightforward mix of the three signals is required. Also, the video signal is not yet encoded and no NTSC-related artifacts have had the opportunity to blur the picture (and therefore the edges) that will eventually be keyed out. As well, the chroma resolution has not been reduced by NTSC encoding.

Internal, external, matte, and chroma keys can all have electronically generated drop-shadows and edges added to the keyed image. The key can also be shown as an outline key, which means that only the outline generated by the edge comparator will be shown; both inside and outside the outline, the background picture will remain.

A very useful, but underutilized effect that has some unique properties is called a nonadditive mix. In a normal dissolve or mix, each source is reduced in intensity—

at the halfway point, each source appears at roughly half its normal video level. In a non-additive mix, whatever source has the highest luminance at that particular point in the picture is the one that is seen. This allows for some very realistic superimpositions of one image over another, providing that the foreground image has sufficient luminance information to overpower the background. The advantage of a non-additive mix over an internal key is that there are no hard edges around the perimeter of the superimposition.

Downstream keys are the same as standard keying processes, except that they are generated by a separate keying module that is "downstream" (electronically) from the rest of the image-creating modules of the switcher. They can be used as a final keying point just before the video leaves the switcher. Downstream key sections of switchers often come equipped with a master fade-to-black bar or auto transition button, to allow the fading out of all video without having to tie up an MLE bus to perform this final effect.

Other Standard Sources

In addition to the sources that we plug into a switcher, there are often other, pre-entered signals. The left-most position on a switcher's buses will be a source of video black. It is used all the time for fade ups, break points, and a position in which the switcher should be left when not in use to prevent video images burning in on monitors. Additionally, there will be a source of colour bars for video recorder and monitor lineup, and as a test signal when making recordings or sending the switcher output to another location.

A background or matte generator is also available to produce a full-field coloured background of any hue, saturation, and luminance desired. Usually, background generators allow gradual shading to black or another colour, or constantly changing (modulated) colour across the screen. Sometimes the background generator will serve double duty as the matte generator for matte keys within the keying module. More often, there will be at least one other separate matte generator reserved for the sole purpose of building these effects. Other places for separate matte generators include border colours on keys, and wipe edges.

Making the Transition

Switchers have the ability to dissolve or wipe between full frame sources, and most have multiple keyers per MLE. To control all of this, a Transition panel incorporates the ability to do one or more of these effects simultaneously, with one press of the AUTO TRANS or CUT buttons, or by using the T-bar. This panel is also usually where the number of auto-transition frames is displayed.

Re-entry and Cascading

So far, we've described a simple switcher with one MLE bus that can do (one at a time) a dissolve, a wipe, or a key. It is possible to make a switcher with more than one MLE bus, and have each bus feed succeeding ones. This arrangement is called a cascading switcher. It is also called a re-entry switcher, since the output of each MLE is re-entered into the input of the next, so effects can be layered on top of one another.

Other Buses

Other useful buses are often found on a switcher. A PREVIEW bus allows you to look at a source before switching to it, as well as set up an effect before taking it to air. In many switchers, the preview function is usually automatically enabled, so the result of your transition (or the transition itself) can be previewed before being taken to air. UTILITY and AUXILIARY buses are used as the name implies. They can be called into service as additional keying sources, or as floor monitor feeds, or for other auxiliary functions.

Electronic Memory

Switchers have the ability to store and recall all or part of the setup that you've so meticulously put together. You can save information on which sources have been selected on which buses, how your keyers have been set up, the colours of your background generators and wipe borders, wipe types, auto transition times, and other related things. There are often several memory banks into which you can store information, for different shows or portions of programs. This is especially useful in quick turnaround situations, such as complicated live-to-air news broadcasts, or for often-used effects in postproduction environments.

Image © Ross Video Limited

Digital Video Switchers

Digital video switchers do all of the same functions as their analog counterparts, but use digital video sources. Cameras, video playback devices, character generators, stillstores, and other devices all send their signals to the switcher in digital format; no analog sources are used. All of the manipulations described above (cuts, dissolves, wipes, keys, and so on) are no longer done by analog video amplifiers, but are instead generated by high-speed computer mathematical calculations on the pixels of these signals. This allows for all of the same effects of analog switchers, but with a high degree of additional manipulation not available with traditional analog methods. As well, the technical quality of these effects is better, since there are no noise or other artifacts from composite NTSC signals to contend with.

Image © Ross Video Limited

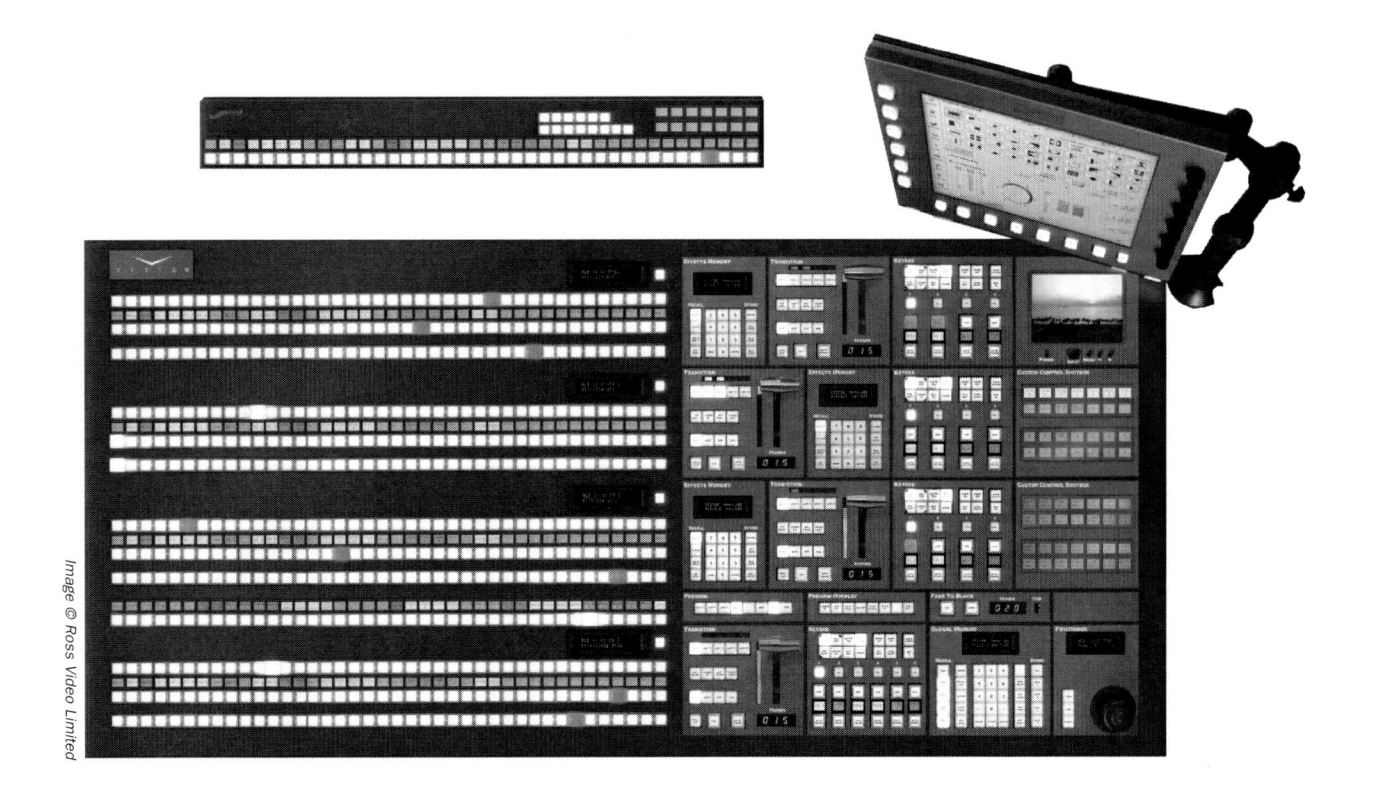

Digital Video Effects

Digital video effects are made possible by devices that change the normal analog video signal into digital information. This information lends itself readily to all sorts of manipulation. Although quite complicated technically, the principle behind digital video effects is relatively simple. The DVE equipment can grab a video frame from any video source, change it into digital information, manipulate it in a variety of ways, and display it. The manipulation information can be stored and retrieved on demand. Digital conversion of analog video can be compared to changing a photograph into a mosaic of the same scene. The photograph shows you a continuous change of colour and brightness. The digital mosaic presents a great number of discrete picture elements (pixels), each one having a solid colour and brightness and its own assigned address within the picture. If you want to change the shape of an object within the mosaic, you can take out some of the pixels, or add some. You can paint onto the picture by changing some of the pixels individually, or in groups, without affecting other areas of the picture. If you remove some of the pixels in a linear organized fashion, you can make the picture smaller; adding some in the same way, enlarges it. All of this manipulation is done very quickly. There are up to 400,000 individual picture elements (pixels) in one frame of standard definition video, and often several pixels will need to be processed just to generate one output pixel. It's not unusual for a modern DVE to perform in excess of 20 billion mathematical operations per second. When digital video effects are interfaced with the standard analog effects of a switcher, the possibilities for visual effects are virtually endless.

Size and shape manipulation in a DVE

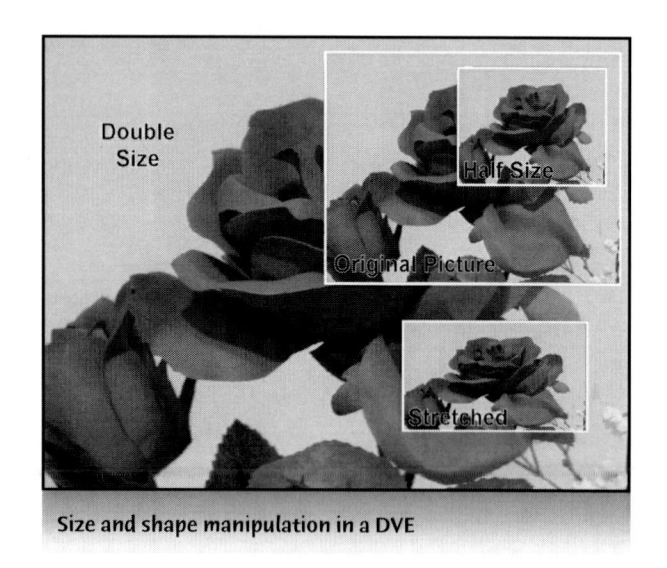

The effect of antialiasing

An almost unlimited variety of effects are available to manipulate the size and the shape of an image. They include compression and expansion, aspect ratio change, positioning and point of view change, perspective, flips (horizontal and vertical), and automatic key tracking.

Compression and expansion involve the basic reducing and enlarging of an existing video picture. It can be the size of a small point, or larger than its original size. If you were to expand a picture to greater than its "times one" size, you would begin to notice that each original pixel became, in fact, a small block—you'd see the mosaic effect on the picture. To reduce this artifact, special mathematical formulas are applied to the picture to average out adjacent pixels, adding more video information at the hard edges between each. This is called antialiasing.

With DVE, you can change the traditional 4:3 (four units of width to three units of height) ratio of standard television pictures to make it anything you like, creating some rather unusual effects. The compressed picture can be positioned anywhere in the frame. It can also be tilted. This looks like the image is placed on a graphic card that has been adjusted so you are looking down it at an angle. The severity of this tilt can be adjusted to make the image quite stretched.

Once the initial effects of size, shape, and positioning had been perfected in the early 1980s, DVE manufacturers sought out even more spectacular effects. This picture manipulation is usually known as warping, and involves effects that turn images into spheres, waves, water droplets, and page turns. As well, images can now be placed onto geometric objects generated by other computers. This process is known as mapping, and it means that surfaces, textures and images can be wrapped around other objects and forms.

The language of DVE motion is the language of cartoons—zoom to here, flip over there, bounce into this. Some of these terms have been coined by DVE manufacturers, others by imaginative (and tongue-tied) production personnel. In essence, motion effects are any of the following: continual changes in picture size and position, zooms, rotations, and bounces. You can pan and tilt a DVE picture, much as you can a camera shot. You can slide a picture off the screen, revealing another. When one picture leaves while another enters, this is a push-off or push-on. Zooms are self-explanatory; notice though, that DVEs can go from very tiny pictures to overblown ones quickly and smoothly. Also, the entire picture moves when you zoom with a DVE; cameras, by comparison, lose the peripheral areas of the picture (or gain them) when zooming in and out.

You can rotate an image on all three axes: the x-axis, representing width; the y-axis, representing height; and the z-axis, representing tilt. In the real world, we think of x-axis rotation as flip (like a flipping coin); y-axis rotation can be considered revolving (like a revolving door); z-axis rotation should be spin (like a steering wheel). Not in the DVE world, though—everybody has their own phrase for these types of motion. DVEs will usually allow you to perform effects on more than one incoming video source at a time.

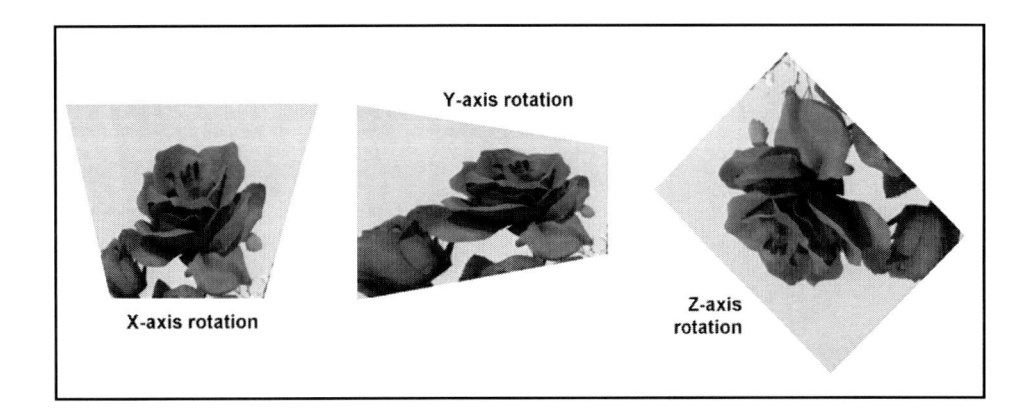

Stillstores

A stillstore is a device used to store electronically many freeze frames for later recall. They usually allow a single field of video or a full frame to be frozen. This is accomplished by locking the video in question into a conventional framestore device. The digital data contained there is then transferred to a computer hard drive, which can hold thousands of images. Selection of a particular image for recall is accomplished by entering a particular frame number, or selecting the image from a listing of available ones. The listing can be in text form, or by way of a series of several small pictures (thumbnails) displayed simultaneously on a video monitor. Once the desired picture is selected, it is quickly recalled from the hard drive and read into the framestore for display and production use.

Generated Graphics

The advantage of electronically generated graphics is that they can be directly integrated into the television system. They eliminate the often time-consuming intermediate steps of preparing a camera graphic and shooting it.

Character Generator

The character generator looks and works like a typewriter, except that it writes letters and numbers on the video screen instead of paper. Before character generators were invented, putting a title or name on the screen involved painstakingly printing, in white ink, paint, or rub-on transfer lettering, the information on a black piece of cardboard called a title card. Character generators can produce letters of various sizes and fonts, and simple graphic displays such as curves and rectangles (blocks). In addition, backgrounds can be colourized. To prepare the titles or graphs, you enter the needed information on a computer-like keyboard. You can then either integrate the information directly into the program in progress, or store it for later use. Each screenful of information is keyed to a specific address, or shown as a thumbnail, for easy and fast recall. Most character generators have two output channels: a preview and a program. The preview channel is for composing titles. The program channel is for actually integrating the titles with the major program material. The preview channel has a cursor that shows you where on the screen the word or sentence will appear. By moving the cursor into various positions, you can centre the information, or move it anywhere on the screen. Various controls allow you to make certain words flash, to roll the whole copy up and down the screen, or to make it crawl sideways. Most character generators can move words and graphics on and off the screen at any angle, and at any of several speeds.

Image © Ross Video Limited

VIRTUAL SETS

From the beginning of television, the bulk of video picture creation has been done with television cameras. In the 1970s, video character generation equipment was invented that relieved us of the tedious task of hand-lettering title cards with people's names and place identification. Shortly thereafter, with the availability of larger computer hard drives, electronic stillstore equipment was developed so that we no longer had to rely on the projection of 35 mm slides within a film telecine, or picture cards shot with a camera, to access nonmoving video material. Modern computer technology now allows us to record and play back moving video from hard drives, and animate, morph, and manipulate video in almost any way imaginable.

We have, over time, tried with varying degrees of success to superimpose live human beings over other video materials. The first attempts (in the days of black-and-white television) were made by shooting talent (who would not be wearing dark colours) in front of a black limbo background. The black was keyed out, and could be replaced with a second video source. This technique was refined in the 1960s with the popularization of colour television, and with it, the ability to choose a particular colour for the keying effect. This allowed talent to wear anything that wasn't the particular hue of the now familiar chroma key background (usually bright green). This method really hasn't changed in over a quarter of a century for television, and has been used since the 1950s in the motion picture industry, where it is called green screen. The traditional chroma key technique requires that the camera and the background information be locked off; that is, there can be no camera movement from either of the two video sources, otherwise the illusion of the person appearing in another environment is lost. Today, thanks to the impressive processing power of parallel connected computers, we have the ability to move the camera in any direction we wish. This is achieved by creating an entirely fictional world within the computer, a virtual set.

First, a virtual set must be designed by the manipulation of three dimensional objects in a computer database. Designers of virtual sets must have studio planning expertise along with 3D modelling and image generation knowledge. They create the new set in a wireframe mode (one in which the edges of the objects in the set are modelled), and then cover the frame with various colours and textures (an operation called texture mapping) until the desired look is achieved. Lighting of the virtual set is also done in the computer, and must be matched later on by identical (or at least complementary) lighting in the green screen studio.

With virtual set technology, camera movement must be tracked in all dimensions, so that the computer will know where and how far to move the virtual set relative to the framing and positioning of the live video. This can be done in a number of ways (or in combination).

The most obvious way of tracking camera movement is to put rotary encoders on every moveable element of a camera mounting: tilt, pan, pedestal (vertical motion), trucking, and dollying. In addition, the camera's zoom, focus, and iris servos must be monitored for changes in framing size, sharpness, and light level. All of these elements of the live shot must be constantly examined, updated, and sent as data to the graphics computer. In earlier virtual set systems, this was done with high reliability and excellent precision.

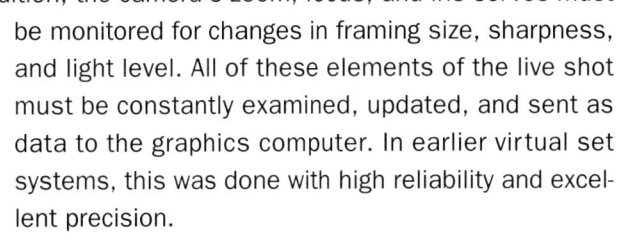

The main disadvantage with the head and servo tracking system is that handheld cameras do not, by definition, fit the tracking paradigm; they have no camera head as a reference point. Other systems have made use of patterns on the chroma key walls or in the grid for relative positioning information. One of the more popular systems for positioning information has been

Orad grid system

developed by Orad. This involves a distinct plaid-like grid system on the studio wall which, because of its nonlinear details, can be recognized by the host computer and can be analyzed for tilt, pan, pedestal, trucking, and dollying information. Another system involves a series of circular patterns mounted on the ceiling grid, and viewed by a small black-and-white CCD camera mounted vertically on the camera which is doing the actual live video shooting. In addition, either of these pattern systems can be augmented by infrared LEDs mounted on the shooting camera, the on-air talent, or both. These beacons, in turn, are viewed and identified by black-and-white cameras located around the periphery of the studio. All of these systems (in varying combinations or by themselves) allow recognition of a unique position within the studio.

The insertion of live performers into a virtual set is done with the same technology as that used in chroma key—the removal of a particular hue of blue or green, and its replacement with the computer generated 3D world. Usually, a more sophisticated chroma removal technique is used, which allows not only fine-tuning of the blue aspects of the key, but also sensing and manipulation of luminance, to create a more realistic superimposition. It's possible to have multiple cameras in a virtual set environment, although the interswitching of foregrounds and backgrounds understandably becomes more complex.

Because of the time it takes to constantly update and render the positioning and movement in a 3D virtual set, there is a several frame delay between the movement of the camera and the final resting position of the updated set. This must be compensated for by introducing a digital delay unit between the camera output and the keyer that will do the final superimposition. On-air talent, of course, will notice that if they go through a certain motion (blinking, a sweep of their hand), a fraction of a second later the same action will occur on the studio monitors. Talent microphones and other associated audio must also go through the same digital delay process to maintain synchronization with the video.

Virtual sets, applied creatively, can be used in a range of applications. Virtual newscast sets have become increasingly common. Virtual sets also allow the viewer to be taken into universes otherwise unattainable, for example, inside the workings of an internal combustion engine, or into outer space. This technology is thus very useful for industrial or corporate training video. Children love imaginative worlds and places, so this is also a logical extension of creative children's programming. Other real-time effects, subtly used, can be just as effective: virtual monitors that drop down from the grid, revealing additional video information; walls or ceilings that remove themselves to reveal nature or the sky; and surfaces that morph into various conventional or unconventional skins: bricks, wood, or waterfalls.

Motion Capture (Mocap)

If we want to animate a virtual character, we can go through the design of that individual, and painstakingly enter in movement and size coordinates

Motion capture actor

for every frame of our animation. On the other hand, it is possible to move the character around, in real time, using a system called motion capture.

In this process, real people wear a special suit on which are attached many reflective spheres. The actors are then placed on a stage surrounded by infrared lights and cameras. Computers pick up the movement of the people via the infrared lights reflecting off the joints of the actors, and translate that into the movement of the 3D character. These movements can be generated live, or the data can be saved for later fine-tuning and manipulation. The result is action characters that have an uncannily realistic movement. This process is often used in videogame design and CGI (computer generated imagery) for television programs and motion pictures.

Infrared lights and pickup cameras

Frozen Action Movement: Time Slicing

Time slicing is an effect where everything appears to be frozen, or in very slow motion, while the camera dollies around the scene. This can often be accomplished with CGI processes, but other methods also exist. One older, but very effective, technique is to record the actor in a green screen environment, surrounded by a series of still photography cameras. The cameras can either be triggered to expose all frames simultaneously, resulting in a completely frozen, but very rotatable moment in time, or sequentially, resulting in a very slow motion, rotating scene. A variation on this method is where the motion picture film is passed sequentially through an array of lenses, each focused on slightly different versions of the same scene.

12

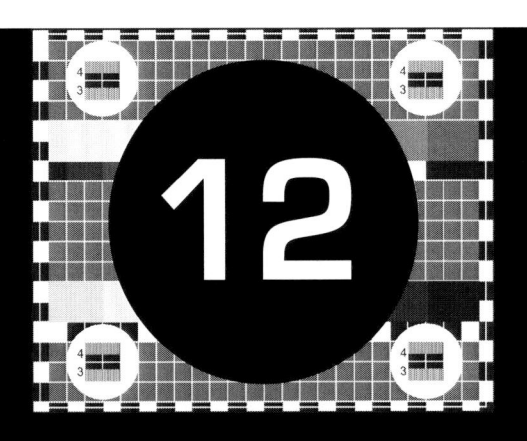

VIDEO RECORDING AND REPRODUCING

"I hate television. I hate it as much as peanuts. But I can't stop eating peanuts."

—Orson Welles, quoted in the *New York Herald Tribune*, 1956

Analog Video Recording

Television's first decade was the era of live TV. Most original programming was broadcast when it was produced. This was not by choice: no satisfactory recording method except photographic film was available at the time. Many live shows were actually televised twice—once for each of the east and west coast time zones.

It was possible to make a copy of a live show by aiming a 16mm film camera at a broadcast quality television monitor and making a recording of the program. This was called a kinescope recording. These recordings were frequently used by stations beyond the reach of cable or microwave interconnections with the networks. The films could be shipped to a local television station, and run through a projector aimed at a special television camera and converted back to video. Because kinescope recordings were expensive and of rather poor quality, almost from the start of television, engineers searched for a way to record video images on magnetic tape.

By the 1950s, magnetic audio tape recordings were widely used on radio, and tape recorders were becoming popular on the consumer market. It was logical, therefore, that if signals representing audio could be stored on tape, their video counterparts also could be recorded. The problem was that the video signal was far more complex, and the frequencies involved were much higher than in audio.

Analog Videotape Recording Principles

Getting the Signal On Tape

Head-to-Tape Speed

There is a problem when we try to record baseband analog video using a method similar to that used in audio. Basic physics states that the highest frequency you can record on magnetic tape is determined by two things: how fast the tape travels by the head, and how narrow the head gap is (the space between the two poles on a recording head).

The formula is: $F_{max} = V_{tape} / 2 \times W_{gap}$. This means that the maximum frequency you can record on magnetic tape is equal to the velocity of the tape passing by the head, divided by twice the width of the head gap.

In audio, the highest frequency that has to be recorded is about 20,000 cycles per second. This means that, at a tape speed of 15 inches per second, you'll need a head gap of .000375 of an inch (20,000 = 15/2 × .000375). This was totally manageable in the 1950s, and made for excellent quality audio recordings. In fact, head gaps could be made as small as .00006 inches (about 1/16,000 of an inch

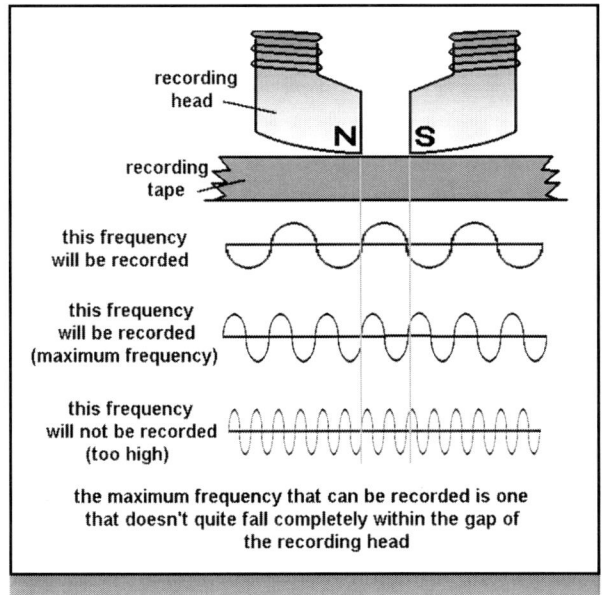

Why there is a limit to how high a frequency we can record on magnetic tape

wide, or about .0015 mm), meaning audio recordings as slow as 3.75 inches per second sounded fine. (Note: We are using mostly Imperial measurement in these equations, since much of the research and development of a practical videotape machine was done in the United States).

The analog video signal has frequencies up to 4.2 million Hz in it. Using our formula, we plug in the numbers, and come up with a tape speed of 504 inches per second (4,200,000 × 2 × .00006 = 504). A half hour recording at this speed would require 75,600 feet of videotape (a little more than 14 miles, or about 23 km.) This isn't a practical solution. But we tried anyway:

1951: The Electronics Division of Bing Crosby Enterprises demonstrated a black-and-white videotape recorder with twelve fixed recording heads. Using high speed head switching, 1" wide tape and 100 inches per second tape speed, a 1.7 MHz recording was made on reels that held 16 minutes of program material—and 8000 feet of tape.

1952: The BBC showed off VERA (Vision Electronic Recording Apparatus), which used two video tracks, tape speeds of 200 inches per second, and which played back at quite low resolution.

1953: RCA displayed their black-and-white system—one continuous video track, with a 17" reel of 1/4" wide tape, moving at 360 inches per second. Later in the same month, RCA proved that they could perform the same feat in colour. They used a 17" reel of 1/2" wide tape, three video tracks (for R, G, and B) along with a sync pulse track, still running at 360 inches per second, and a recording time of . . . four minutes.

How the Quad Did It

The breakthrough finally came in 1956 from a team of engineers working at Ampex Corporation in California. They thought out of the box by realizing that there was no reason that the video recording heads had to be rigidly fixed to the recording transport. What if they were free to move across the tape at a very high speed, while the tape was pulled past them at a much slower speed? They solved the problem by mounting the recording heads on a wheel that revolved at 14,400 rpm, impressing the signal crosswise on the tape. As a result, the actual head-to-tape speed was very high, but the tape transport moved the reels of tape at a relatively slow speed, 15 inches per second. This invention was the quadruplex videotape recorder, colloquially called the "quad," because of its use of four recording heads. It used tape 2 inches wide.

How the Quad Works (and why we care...)

Here's a little background on what the folks at Ampex designed into their machine.

In one second, the head drum rotates 240 times, placing 960 short tracks on the videotape. The tape is 2" wide, but room is left for audio, control track, and blank, separating guard bands between all of the tracks. Therefore, each video track is about 1.8 inches long. To allow uninterrupted output from the switched heads, only 1.625 inches is used for fresh video; the rest is overlap from the previous track's video.

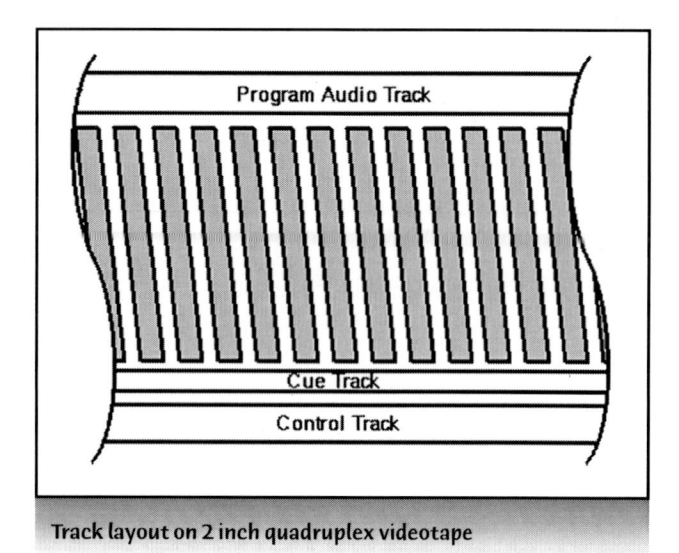

If we do the math on the Ampex system, we find out that the effective head-to-tape speed is 1560 inches per second (1.625 × 960 = 1560.) Using our familiar "maximum frequency we can record" formula from earlier in this chapter [$F_{max} = V_{tape}/2 \times W_{gap}$], this means we can record a frequency of up to 13 MHz—lots of room for our 4.2 MHz video signal (1560/2 × .00006 = 13,000,000 Hz.) It's not quite that simple, however.

Track layout on 2 inch quadruplex videotape

Linear Recording and Playback Frequency Response

The next problem is one of the strength of a signal relative to its frequency. If we try to record a full-range analog video signal directly on tape, we find out that the higher the frequency of the recorded signal, the greater the playback output from the tape. Conversely, the lower the frequency, the lower the playback output.

This is a big problem, because we are trying to put on videotape a signal between 30 Hz and 4,200,000 Hz. This is a 140,000:1 ratio. And that, in turn, means that the playback voltage at 30 Hz will be 1/140,000 as strong as that at 4.2 MHz. Expressed in dB (as we do in audio), this ends up being a range from the lowest to the highest voltage of about 103 dB. That's like trying to record a quiet conversation and an airport runway at the same time. This is not good.

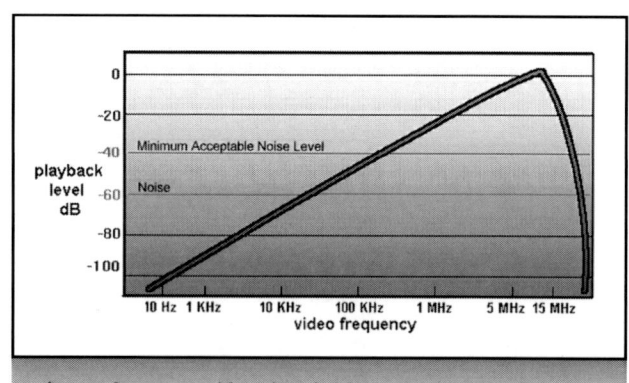

Why we don't record baseband video onto the tape directly

What are the ways around this problem? Using a super equalizer to hold down the more powerful high frequencies and boost the less powerful low frequencies (thereby making the frequencies more "even" to one another) is one idea. Unfortunately, in 1956, most analog amplifiers and equalizers had electronic noise in them below about 60 dB, due to the nature of vacuum tube electronics systems and the early transistorized technology of the day. At any rate, that's still a long way from our 103 dB required range. So, using an expensive and sophisticated equalizing method won't work because all of the lower frequencies would still be hopelessly lost in the noise of the amplifiers themselves.

The solution lies in a device called a modulator. The word "modulate" simply means "change" and that's what a modulator does; it changes one parameter of an electrical signal, using a second electrical voltage to tell the modulator what to do. The first signal, called a carrier, is sent into the modulator. The second signal, the modulating one, varies a parameter of the carrier, either its level (or "amplitude") or its frequency. The output is either an "amplitude modulation" or a "frequency modulation" of the input signal—also known as AM and FM.

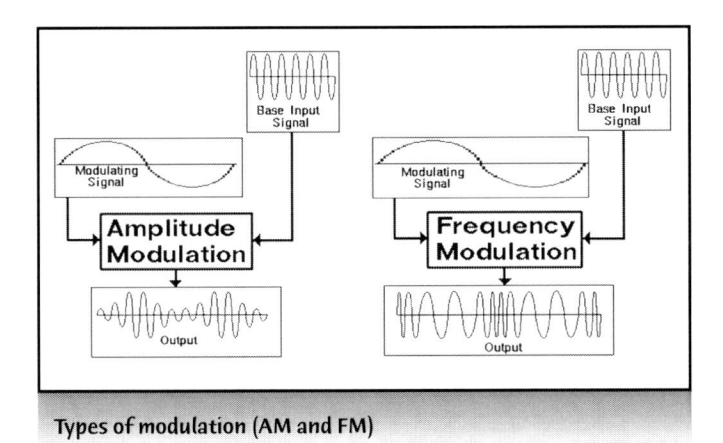

The quadruplex analog videotape recorder uses frequency modulation. A constant frequency carrier of 8.6 MHz is modulated by the incoming video (which may be between 30 Hz and 4.2 MHz). The modulator output will therefore vary the frequency of the original carrier from 4.4 MHz to 12.8 MHz (8.6 MHz − 4.2 MHz = 4.4 MHz; 8.6 MHz + 4.2 MHz = 12.8 MHz). Within these frequency modulated fluctuations is a replica of the program video, just encoded in a different way than the original analog video signal.

Types of modulation (AM and FM)

What's important to us is the ratio of the highest to lowest frequencies on the tape. It's now about 3:1 (12.8 : 4.4). This is a little less than a 10 dB difference. The noise in any electronic amplifiers we use will be back at −60 dB, which is an excellent safety margin. To use this new modulated signal, we'll need to be able to record up to 13 MHz of information on our VTR. With the quad system, and its effective head-to-tape speed of 1560 inches, we have that recording capacity.

When playing back the modulated video, it will have to be converted back to analog video again. A device called the demodulator has the reverse function of the modulator: it takes the high-frequency signal from the tape and retrieves the video signal from it. This video signal is then put through various processing circuitry to make it stable and to adjust its levels, at which point it looks like the video signal that was originally sent to the videotape recorder.

Servos and Timebase Correction

Keeping Track of the Tracks

When a videotape is recorded, there must be a way of letting the videotape recorder know, later on, when and where the video tracks were laid down, so they may be played back reliably. In addition, the motor spinning our rotating video heads (the head drum motor) and the motor that pulls the videotape through the transport (the capstan motor) must be kept running at a very constant, identifiable speed, referenced to the incoming video. We perform these functions with servos.

Drum and Capstan Servos

A servo, in general terms, is an automatic control mechanism in which the output of a device is compared with a stable input reference, so corrections can be made. To understand what a servo is, and what it does, consider the following hypothetical predicament. You've been chosen to keep the speed of an electric fan motor constant. To do that, we've supplied you with some equipment: a control knob to speed up and slow down the fan, and a little counter that's attached to the fan's motor shaft. Every second, the counter starts counting the revolutions of the fan motor. Your job is to keep the counter always at 100 revolutions per second—no more, no less. In the first second, the counter counts up to 102; you turn the speed control down a little bit, to slow down the fan. In the next second, the counter only counts up to 97 (you went too far), so you turn up the knob a little bit. In the next second, the counter is up to 99; you turn it up a tiny bit more. The subsequent count says you're up to 101, so you back off the knob a little bit. And so on.

You're a servo. You're continually comparing the speed of the fan to a constant (100), and adjusting the speed to try to match it. In a videotape recorder, electronic circuits replace the controller knob, the little counter display, and your hand. Everything is done electronically. That's an electronic servo.

A human servo

The Drum Servo

The drum servo's function is to try to keep up a constant speed for the head drum, as its video heads move across the videotape. During record, it looks at the incoming video's horizontal sync pulses and ensures that the VTR lays down the proper tracks for each frame of video. During playback, the drum servo can lock either to the playback video itself or to an external reference from a central sync generator (within the broadcast facility) to ensure that it plays back the correct video tracks to reproduce the video.

The Capstan Servo

The capstan servo, on the other hand, is trying to ensure that the tape is being pulled through the videotape recorder at a very constant speed. During record, the capstan motor runs at a certain speed, which is compared to the timing of the incoming video's vertical sync pulses. While this is going on, a control track is also being laid down on the videotape. This is a pulse for every frame of video, about 30 pulses a second. On playback, the capstan servo looks at the control track and makes sure that the tape is being played back at exactly the right speed, reading 30 pulses per second from the tape. In broadcast VTRs, this is often referenced to the station's external vertical sync pulses.

The control track signal is also used for setting up the videotape's position relative to the video head drum, so that it can reliably pick up the tracks recorded on the tape. The capstan's speed is adjusted during the "lock up" period just after pressing

Chapter 12 Video Recording and Reproducing ■ **179**

Videotape recording basics

Videotape playback basics

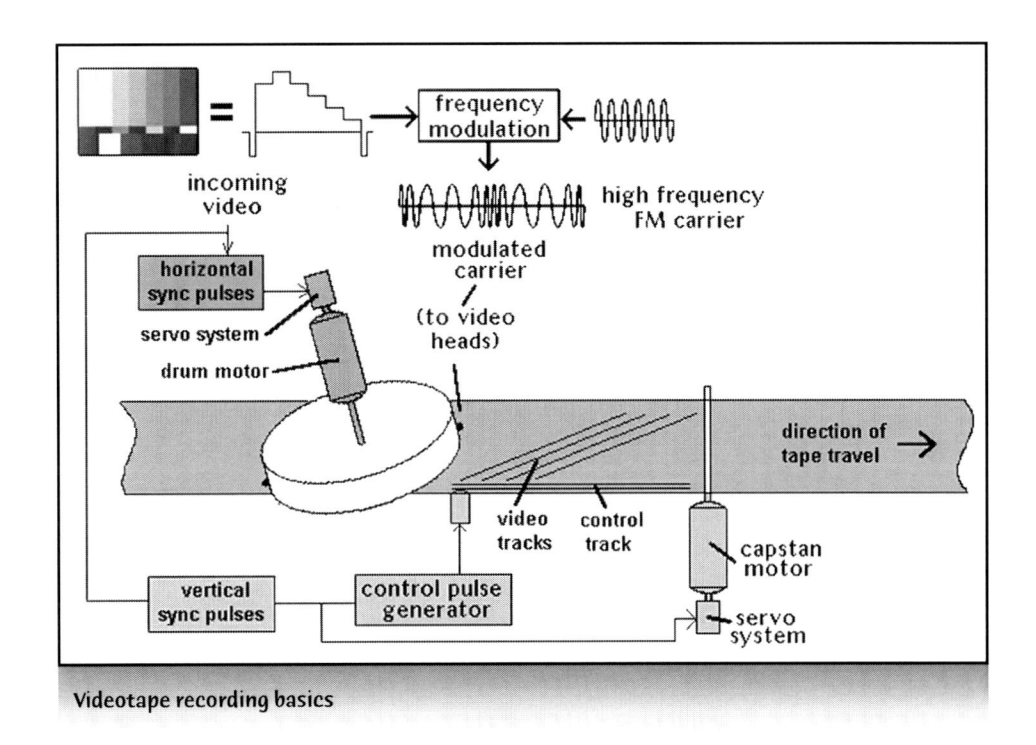

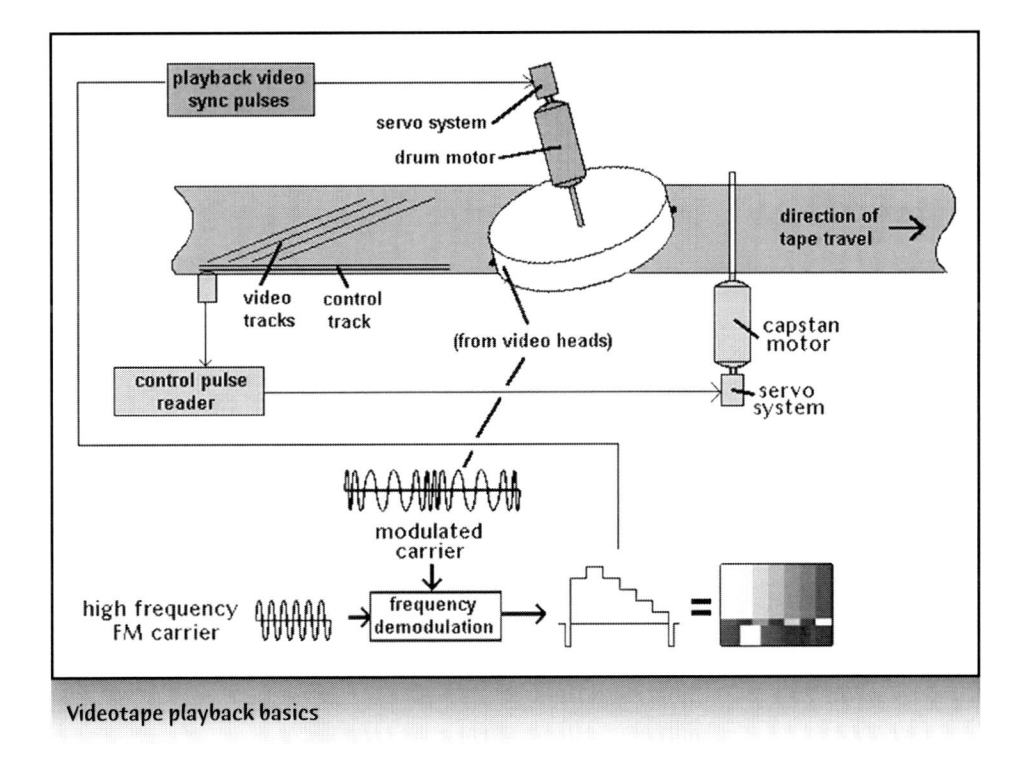

the Play button, so the video tracks on the tape line up precisely with the video heads sweeping by. This process is known as "tracking," and it usually takes a second or so for all of the systems to stabilize properly. That's why, when you push the Play button on a VTR, the picture is unstable for a moment: the VTR's servos are "getting their act together." And that's why, in a production environment, we roll a VTR a few seconds before we need to take it to air—to allow for the VTR lock-up time.

Timebase Correction

All VTRs are mechanical devices, and there is a limit to the timing accuracy that can be achieved on playback. The limit is set by the mechanical accuracy of headwheel bearings and motors, and a certain minimum variation in tape quality along the length and width of the tape. Also, gradual and sudden variations in the position of the tape guides inside the VTR, and the tape path itself during extended use, may cause instability. Timebase correctors are used to take out the remaining mechanical jitters in videotape playback so that the signal is stable enough for broadcast use.

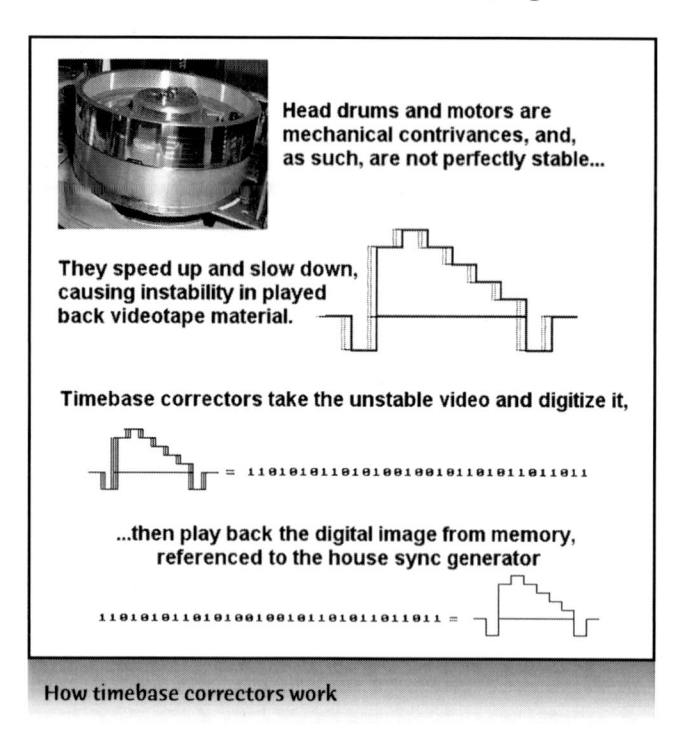

The first quad VTRs had two devices within them. The AMTEC (Ampex Tape Error Corrector), the first analog timebase corrector, was essentially a series of electronically variable video signal delay lines used to compensate for the playback timebase error. As the head drum speed deviations were detected, a signal representing these errors would be sent to the AMTEC, and the video signal would be delayed as necessary to compensate for the instability. The COLORTEC was later designed for less major errors, in order to provide a stable colour playback signal.

Today, modern timebase correctors not only allow us to compensate for mechanical timebase errors, but, since they're digitizing the video anyway (usually a frame at a time), they also allow us to adjust video, black, and chroma levels, and hue. This correction technology is used not only in composite video, but for component video formats and digital videotape recorders as well.

Head drums and motors are mechanical contrivances, and, as such, are not perfectly stable...

They speed up and slow down, causing instability in played back videotape material.

Timebase correctors take the unstable video and digitize it,

= 11010101101010010010110101101101

...then play back the digital image from memory, referenced to the house sync generator

11010101101010010010110101101101 =

How timebase correctors work

Analog VTR Formats

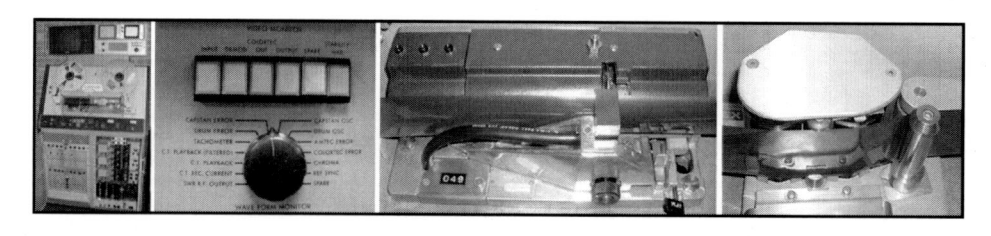

Because the quad was the venerable workhorse of television videotape for the first twenty years, we've been concentrating our understanding of analog videotape recording theory on a very outdated videotape format. But there's more to it than that. The quadruplex videotape recorder embodied all the principles used in modern helical scan machines (which replaced the quad starting in the late 1970s), but in ways that are easy for all to see. Indeed, each component and section of a quad can be literally physically removed from the machine itself, and studied further. By contrast, modern VTRs are often highly centralized marvels of construction; often more than one subsystem will be on a single printed circuit board, or even within one integrated circuit chip.

While much has been said for the hands-on approach to learning about videotape theory of operation, the fact is that so much of this background information is now seldom used because of the comparable high reliability of today's equipment, and the relative lack of operator adjustments. There are hardly any lineup controls on the front panels of such equipment, except input and output level controls and tracking adjustments. Even these can often be placed in automatic or preset positions.

With this in mind, it is now time to take a brief look at the myriad formats and features that have been available over the years. The following list, while not exhaustive, is nevertheless historical—professional analog videotape machines have not been manufactured for some years now. Readers who are interested in a full list of videotape recording formats (both analog and digital) are encouraged to look on the Internet, where there are many authoritative sources of information.

- **1956—Quadruplex**—Developed by Ampex Corporation, this was the first viable videotape recording system. Originally designed for black-and-white television, it was upgraded in the early 1960s to record colour programs. It was still used up until the early 1980s, while being gradually replaced by the Type C 1" videotape recorder. The quad system was also used in early videotape cartridge machines, such as the ACR-25, allowing short duration commercials to be played back in quick succession.

- **1967—IVC 700/800/900**—An early 1" videotape format, used in some television stations because of its lower cost. Unfortunately, its quality did not match existing quadruplex videotape machines.

- **1970—3/4" U-Matic**—Initially introduced as a videocassette home recording format in the early 1970s (consumers found it too big and clunky, with a short recording time), this was the first helical format to be widely accepted and was ultimately to find its niche in the ENG (electronic news gathering) and industrial video markets.

- **1976—Type C, 1"**—Jointly developed by Sony and Ampex, this was the first machine to add special effect capabilities (e.g., still frame, slow motion, and the ability to see pictures while shuttling the tape at high speed) that were not possible with the 2" quad VTR. This machine was the North American television industry's workhorse from the early 1980s to the early 1990s, as it had a quality rivaling the legacy 2" videotape machines then in use.

- **1976—Betamax**—Not to be confused with Betacam (the broadcast format, described later), this was the first standardized consumer video format and used a composite recording system. Later Betamax versions introduced the concept of "HiFi" audio using an FM subcarrier.

- **1976—VHS**—This system came out at the same time as the Betamax system and, despite its poorer picture quality, outsold Betamax due to longer playing times on the cassettes. The "HiFi" version of this machine used a recording system for its audio in which a long wavelength FM carrier was recorded under the high-frequency FM video carrier. VHS recorders are still being manufactured, if only because consumers have vast legacy libraries of television programs and home movies in this format. S-VHS, a system designed as a high-end consumer format also used in industrial video markets, used a high coercivity tape, and the maximum video FM carrier frequency was increased, resulting in picture resolution improvement. With the proliferation of inexpensive and compact MiniDV digital camcorder and playback VTRs, S-VHS eventually became an obsolete format.

- **1982—Betacam**—Betacam was the first of the small format machines intended for ENG (electronic news gathering) and EFP (electronic field production) use. Its use of an upgraded consumer format cassette (Betamax) was the notable feature, as well as its being the first professional component recorder (the luminance and chroma signals were recorded on separate tracks allowing each of the two signal channels to be optimized for their particular requirements). The tape ran six times as fast as consumer Betamax machines. Many stations found that Betacam was perfectly useable for general purpose on-air use, including program and commercial playbacks. The overall look was very slightly noisy because of the amount of signal processing going on, but probably below the threshold of visibility for most people. When used in a completely component environment, the results were very good. Betacam's close relation, Betacam SP, used a metal type tape for higher quality video reproduction. There are quite a number of these machines still in use today, largely because of the fact that many television stations have large libraries of Betacam footage.

- **1983—8mm, Video 8, and Hi8**—The 8mm format was the first time a consumer video machine made use of an imbedded tracking servo rather than a longitudinal control track. This allowed the use of a very slow tape speed, metal evaporated tape, and a small cassette size. You'll also notice that there is no tracking knob on any 8mm format machine. HiFi audio was available through an FM subcarrier recording system, with PCM (pulse code modulation) digital audio as an option built into the format specifications. For consumer camcorder work it was ideal, with one of the best format size to picture quality ratios, for an older analog format. It continues to be used as a consumer camcorder format, and tapes can still be purchased.

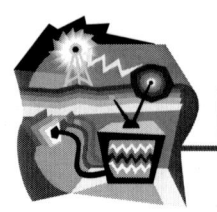

DIGITAL VIDEO RECORDING

Videotape recording gave broadcasters a valuable tool. As each new analog format appeared, signal processor and editor packages emerged to stabilize its electromechanical irregularities. Yet, an undesirable multiple-generation syndrome continued to surface when editing in traditional tape environments, where the editing process involved essentially making a dub, or copy, of the camera tape onto a second, master edited tape. What we call video noise is in reality a combination of phase modulation, transport jitter, amplifier noise, improper machine lineup, and a host of other small inconsistencies. Digital videotape recording tried to solve these problems.

The physical reality of magnetic recording is ideally suited to digital concepts. Information is stored on tape according to the polarization of the magnetic particles in the oxide. In making an excellent saturated recording, all particles will be magnetized completely in either the "north" or "south" direction, exactly what we want for a digital recording (the "1"s and "0"s of the digital domain). In addition, error correction and concealment alleviate the effects of tape irregularities and events that leave momentary gaps in the recording; this is achieved using some elegant error correction and error concealment procedures. Analog recording places adjacent parts of images next to each other on the tape. The concept in digital recorders is to spread pieces of the image over a large area on the tape, so no adjacent pixels are written side by side.

DIGITAL VTR FORMATS

What follows is a non-exhaustive listing of some of the major digital video recording systems that have been available. However, as camera recording moves to digital memory chip recording technology, and as editing is now largely computer-file-based, the list of digital tape recorder formats will likely remain more or less static, not to mention eventually becoming a historical footnote, although readers will still find some of these formats in the field for a few years to come.

- **1987—D1**—This was the first digital videotape format, and was a component format. The digital video was not compressed. D1 was based on the "4:2:2" component system described in CCIR Recommendation 601, which we have discussed earlier. This format was well suited to original high-quality non-NTSC digitally sourced work, such as graphics and film transfer. D1 used a 19mm metal-particle tape that was available in three different sizes of cassettes; the large one held 96 minutes of program material.

- **1986—D2**—This was an NTSC composite format. This format was well suited as a workhorse in the broadcast environment, for handling program and commercial playbacks, as well as satellite recordings, since all of it was received by the station as NTSC video. D2 was designed to be a direct plug-in replacement for analog 1-inch and Betacam recorders, whereas D1 needed to work in a completely non-NTSC environment. D2 operated at a 4× subcarrier (14.31818 MHz) sampling rate. While all of this looks good as a theory, the fact of the matter is that D2 never really caught on, due to its initial upgrade cost, and that it was quickly supplanted by D3 and later digital recorders.

- **1991—D3**—This format, designed by Panasonic, used 1/2" tape instead of the 19mm (3/4") tape used in D1 and D2. It also used a different error correction system, a different modulation coding system (for recording), and a low-tension, low-friction transport. D3 was plug-in compatible with D2 since it was a composite format. In addition, D3 used a flying erase head for insert editing. D2, in contrast, just recorded over previous material, which, in fact, worked out because of the nature of digital recording.

- **1992—DCT 700d**—One of the last systems designed by Ampex, it used 2:1 compression on a proprietary tape format, 3/4" wide. It was an 8-bit format, taking composite, component, and digital component inputs and outputs.

- **1993—Digital Betacam**—Developed by Sony, this was a 4:2:2 component VTR with 10-bit resolution and a 2.34:1 compression ratio on a 1/2" metal-particle tape. This VTR also played back legacy libraries of Betacam SP tapes, making it a valuable addition to then-developing analog/digital hybrid videotape editing systems.

- **1994—D5**—This machine was D3 playback compatible, but offered 10-bit processing (which some say improved the quality of the recording), and could be used as either a composite or component VTR.

- **1995—D6**—This was a digital HDTV recorder that recorded 64 minutes of 1.2 GBytes/sec video on a D1 large-shell cassette. Luminance sampling was 72 MHz and chroma sampling was 36 MHz. Sampling happened on an 8-bit level. Notice that this format specified its recording ability not as NTSC or HDTV per se, but as a capability of recording so many bits per second.

- **1995—D7**—This was the standard for Panasonic and DVCPRO, which was a 4:1:1 component machine with 8-bit recording and 5:1 compression using DCT (direct cosine transform). Notice that this was the first "D" series machine to use compression. This format was also compatible with the many MiniDV camcorders that were in use.

- **1995—D9**—This was the proposed standard for Digital-S, the format from JVC that was to replace S-VHS. Digital-S used full 4:2:2 sampling, with 8-bit recording and a 3.3:1 DCT compression scheme, in a 1/2" tape in a shell that looked like an S-VHS cassette. The specification called for four channels of audio. Its video inputs accepted composite, component, digital component, digital composite, and S-video. The machine would also play back existing S-VHS libraries. This format never really caught on, though.

- **1996—Betacam SX**—Also developed by Sony, this was a 4:2:2 component VTR with 8-bit resolution and MPEG-2 compression of 10:1, recording on 1/2" metal-particle tape with four channels of 16-bit audio. This VTR also played back legacy libraries of Betacam SP tapes, such as Digital Betacam.

- **1996—DVCAM**—Yet another format developed by Sony, it was a 4:1:1 component system, 8-bit resolution, with 5:1 DCT intraframe compression. It was similar to D7 (DVCPRO) in many respects, but was incompatible. It would play back Sony's consumer DV tapes, making it possible to record on inexpensive digital camcorders for shooting in more "adventurous" situations.

- **1997—D11**—Also known as HDCAM, it was a format that used 1/2" metal-particle tape (unlike D6, the other HDTV format, which used 3/4" tape.) The idea behind this was to enable handheld HDTV recording, because of the smaller transport. There were some tricks to this, though. HDTV was designed with a resolution of 1,920 pixels across the screen; HDCAM reduced this to 1,440 (4:3 subsampling). This was represented, in the shorthand of the day, as 3:1:1.

- **1998—DVCPRO 50**—This was a close cousin of DVCPRO, but was a 4:2:2 component recording system, instead of 4:1:1. Its compression rate was a little less, too, at 3.3:1. It would take composite input as well as component.

- **2000—DVCPRO 100**—Also known as DVCPRO HD or D12, this was the HD version of regular DVCPRO. This was the format used by Panasonic in their HD cameras, and generally recorded on solid-state P2 cards, not videotape.

- **2001—D10**—Also known as MPEG IMX, this was a digital component system, which would record and play back MPEG encoded video. It would also play back all Betacam formats (digital or analog) except for HDCAM, making it a possibly excellent replacement for existing workhorse VTR playback and record applications such as master control, satellite feedroom recordings, and so on.

How Digital VTRs Record and Play Back

The digital systems vary somewhat in their processes and track layouts on the videotape, but the principles are the same for them all. In some systems, the video enters the recorder as digital information or component video; in others, the video is in composite form. If the video is in anything other than digital form, it is first converted to digital via A/D converters. Once the video is in digital form, it is pre-coded—a kind of organized scrambling. This effectively breaks up the video into a pseudorandom pattern; originally adjacent pixels are no longer next to each other. Similar procedures are performed on all of the analog audio channels as well.

Because of all the shuffling and coding that's taken place, it's possible that a very low frequency signal has been created (one without a lot of fast changes in the "0" and "1" pattern). This signal cannot be recorded to tape; you need voltage fluctuations to make a recording. The solution has its direction in mathematical and physical theory that states that random noise has no very low frequency (DC) component. Therefore, the shuffled data stream is modulated (digitally, this time) with a pseudorandom sequence of "1"s and "0"s, called "non return to zero transition on I" (NRZI) code.

Damage Control

Suppose that you lose 20 pieces out of a 1000 piece picture puzzle before it is assembled. The result is not total destruction; when the puzzle is assembled, the error is spread out. If you view the assembled picture, you can still see it, although you may need to interpolate some spots. That's the concept behind processing in the digital formats. In the digital recording and playback system, mapping transforms are used. Not only is a field broken into segments and sectors, but also the segments, sectors, and individual lines undergo shuffling. As a result, tape damage, instead of destroying portions of the picture, may appear as little faults spread over the picture.

Error Correction

Here's an example of how error correction can work. Suppose you have the following values of three pixels: 3, 7, and 10. If you add the values together, you get 20. This total is called a checksum. You can make an additional checksum by weighting the values of the pixels and adding them up again. If we multiply pixel #1 by 1 we get 3; multiplying pixel #2 by 2 gives us 14; multiplying pixel #3 by 3 gives us 30. If we add these three new numbers up we get 3 + 14 + 30 = 47. This is our second checksum. Notice how the weighting factor is dependent upon the position of the pixel.

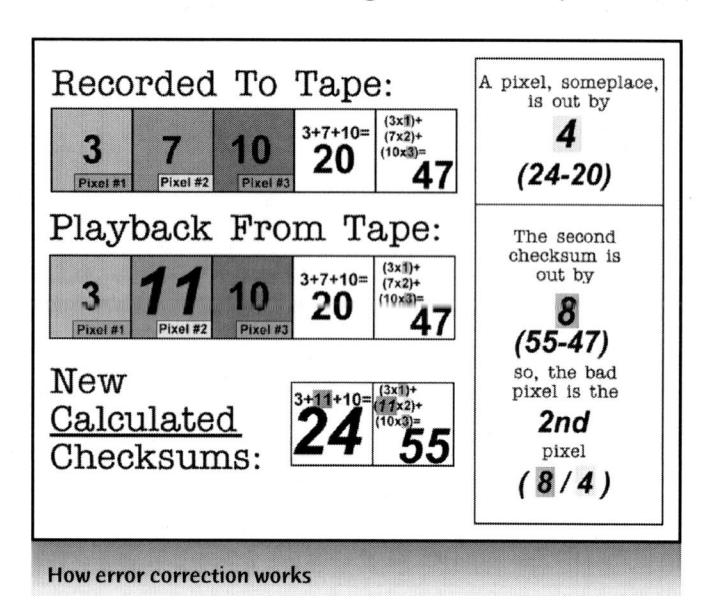

Suppose that now we read our three pixels off tape, and we get the erroneous values of 3, 11, and 10 (the second pixel, in this case, is wrong). The addition of these values comes to 24; the addition of the weighted off-tape values (3, 22, 30) comes to 55.

Using our original checksums, and the off-tape incorrect checksums, we see that we're high by 4 (24 − 20) in our values some place, and it is in pixel number two (55 − 4/24 − 20). Sure enough, it's the 7 pixel that got read out as an 11. We can correct it, and replace it in the picture with the proper value.

Error Concealment

The other process used to suppress problems is error concealment, which is a form of interpolation. In dropout compensation in an analog VTR, missing information is replaced from the previous line's video. In error concealment, values of a group of adjacent pixels are applied to a mathematical formula to determine the probable value of the missing pixel. These algorithms can be quite complex. If minute details involved many single, separated pixels of vastly different luminance and chrominance, and a pixel was missing somewhere in this complicated video detail, error concealment might not regenerate the lost data properly. Television, however, seldom has this type of detail, especially over a period of several TV fields.

Imagine a group of nine adjacent pixels, with all of the values known except for the centre one (see the accompanying illustration.) The simplest interpolation of a situation like this is to just add up all the outer pixels and divide by 8:

$$7 + 8 + 7 + 6 + 7 + 5 + 6 + 5 = 51 \div 8 = 6.375$$

The missing pixel has a value of about 6. This is a simple algorithm, but it illustrates the principle.

Track Size

With the frequent advancement in digital and analog videotape recording, there has been a parallel reduction in the track width. The moral of this story is simple: keep your videotapes and videotape machines scrupulously clean, and always store videotape in a protected environment, away from extremes of heat, cold, humidity, magnetism, and dust.

DIGITAL VERSATILE DISK (DVD)

A later evolution of the compact disc (CD), the DVD has a storage capacity that is seven times higher than its older relation. The recording pits are less than half the diameter of a CD's, and the track can sometimes exceed eleven kilometres in length. All of this adds up to 4.7 gigabytes of storage—typically room for 135 minutes of widescreen (16:9) video at MPEG-2 quality accompanied by multiple audio and subtitle channels. This is called the DVD-5 format. Two-sided versions are available, too: a flipable format called DVD-10. With this capability, along with multilayer versions of the disk, storage capacity of up to 17 gigabytes is available, the DVD-18.

While not strictly a broadcast video recording format (it's generally a write-once technology) it has its uses as a more or less permanent storage medium where applications require many short duration elements for live or edited production, quick changes in live-to-air elements, or several versions of interstitial elements. In addition, this format is readily accepted into the home due to its relatively low cost and its high quality compared to VHS videotape recorders.

BLU-RAY

With the advent of high definition (HD) television, a standard SD DVD no longer has enough room to store a full-length motion picture. A single-layer Blu-ray disc can store 27 gigabytes of data, which is enough for a two-hour program; the double-layer version holds as much as 50 gigabytes of information. A Blu-ray player's interactive features also allows consumers to download additional information from the Internet (captions, interactive data, new video programming, and so on), to record one program while watching another, to view picture-in-picture content, and to edit or re-order the playback sequence of content on the disc.

Blu-ray discs are recorded using a blue-coloured laser beam (hence their name). Because the wavelength of blue light is shorter than red (the colour normally used in a DVD recorder's laser), the pits recorded in the Blu-ray format are only .15 microns across (a micron is one-thousandth of a millimetre). This is about one-third as wide as a DVD's. The tracks are also spaced about half the distance away from each other, relative to DVD's track width.

13

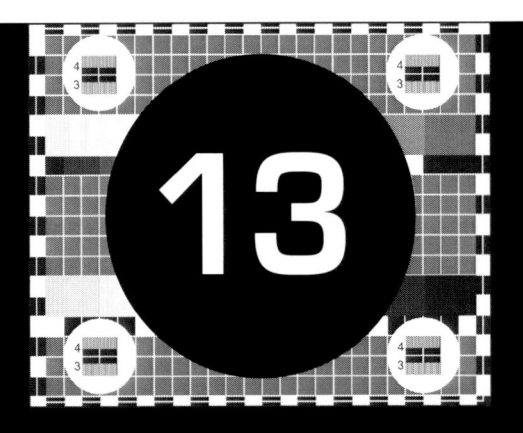

EDITING

"An editor is one who separates the wheat from the chaff and prints the chaff."

—Adlai Stevenson

Why Edit?

We do postproduction work to accomplish the following:

- combine diverse elements
- trim existing elements
- correct mistakes
- build up shows from smaller segments
- execute special effects

The simplest editing is when you combine program portions by simply cutting the various prerecorded pieces together into the proper sequence. The more care that was taken during the preproduction, the less work you have to do in the postproduction stage.

Many editing assignments involve trimming the available material to make the final video fit a given time slot or to cut all extraneous material. This occurs in ENG editing, where you may have 10 minutes' worth of exciting fire footage, but only 20 seconds to tell the story.

Editing is often done to correct mistakes by cutting out the bad parts, possibly replacing them with good ones. This can be quite simple and may only involve cutting out a few seconds during which the talent made a mistake. It also can become quite challenging, especially if the retakes do not quite fit the rest of the recording, because of changes in colour temperature, background sounds, continuity, or field of view.

The most difficult, but most satisfying, editing assignments are those in which you must build a show from a great many takes. In this case, the edit is the major production phase. This is especially true in EFP postproduction, when all takes are shot with a single camera to be combined later.

The major advantage of editing (versus live production) is that you can take time for reviewing the unedited material and deciding where to cut without tying up an expensive multicamera studio environment. Too many times, however, people start editing without having properly thought about the editing sequence. Skipping the planning stage can sometimes help to save time, but more often than not you will get lost in a maze of detail. In all but the most routine editing jobs, you will need to do an editing outline—a list of the desired event sequences and the necessary transitions.

SMPTE TIME CODE

Time code is a way of representing time and position information about a video clip. SMPTE time code is an electronic address for each frame of video. This address is recorded on the time code track of a videotape, or encoded with the digital video on a hard drive or solid-state recording medium.

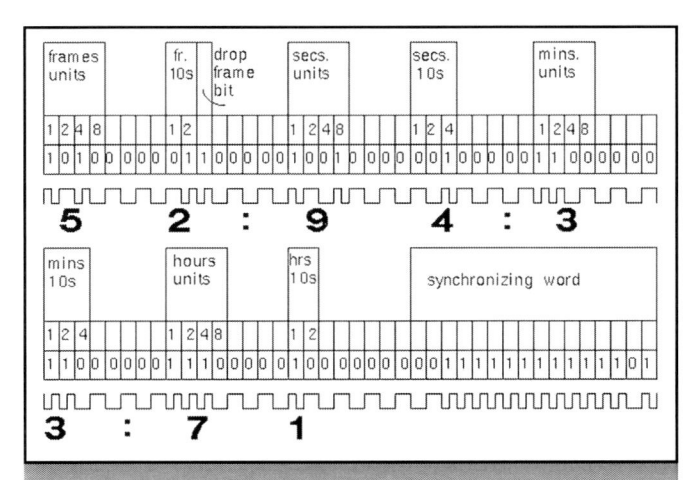

How the time code number 17:33:49:25 is generated using SMPTE time code

The code itself is made up of 80 binary digits (bits) of information.

There are several groups of "assignable" bits for reel and show IDs and the like, as well as some future expansion room in the form of "unassigned" bits. These single digits can be any hex (base 16) value (0–9 and A–F). These can be used as a date, a "scene and take" number, or even as a source identifier.

The remainder make up the "hours:minutes:seconds:frames" we're familiar with and a synchronizing word. This word always contains the same information to provide time code readers with a clue to when each time code word begins and ends and, if using videotape, which direction the tape is moving. The sync word of the time code frame must correspond exactly with the vertical interval of the matching TV frame. To see how a typical number (17:33:49:25) is sent, look at the accompanying illustration.

Usually found on videotapes, vertical interval time code (VITC) is an analog representation of time code. VITC is located in the vertical interval of a signal where it can be read back by equipment capable of doing so. It's essentially a series of white and black patches on one line of video. These segments are interpreted as "1s" and "0s" in the time code reader.

The time code reader, in its simplest form, is a box that takes the VITC representation of time code and converts it into a displayed set of numbers. It can display on an LCD readout or on a "burn-in" within a video signal.

Burn-in of time code on video

What Is Drop-frame Time Code?

Non-drop-frame (or NDF) counts 30 frames per second. However, colour television has, in fact, 29.97 frames per second.

In drop-frame (DF) time code, frame numbers 00 and 01 are dropped from the counter every minute, except multiples of 10 minutes (10, 20, 30, 40, 50, 00).

Therefore, 108 frame numbers are dropped every hour (3.6 seconds), or one frame number about every 33.3 seconds. Notice that it's digits from the numbering system that are dropped, *not* actual frames of video. You still have all of your video information with drop-frame time code.

Analog Editing Systems

A Historical Overview

When videotape was first introduced, tape was actually cut with a razor blade and spliced together. It was important to cut between the individual tracks of video, otherwise jumping or rolling of the video picture would occur. The Smith splicer allowed precision physical edits to be made. To make an edit in those days, you as the editor would push the stop button where you wanted the edit to occur. You'd mark the tape with a grease pencil, carefully unthread the tape from the quad machine, then put it in the Smith splicer. Because you wouldn't be able to see the video tracks directly, you would wipe a "developer fluid" (essentially very fine iron filings suspended in alcohol) over the videotape which would reveal the tracks. Using the microscope in the splicing device, you'd find the nearest control track pulse, and gingerly make your edit with a sharp hinged blade. The second piece of tape (the continuation of your show) would be stopped, marked, developed, and hacked in a similar fashion. Finally, the two pieces would be put together with adhesive splicing tape. If you were good, the picture wouldn't roll or break up (and the join wouldn't fall apart) as each splice went through the high-speed rotating heads with a distinctive "zinging" noise.

Nonlinear Editing

In the beginning, moving images were recorded on motion picture film. Editing film meant finding the right shot on the reels (via a Movieola or flat bed editing machine); cutting on the individual frame lines; and splicing using tape, film cement, or a hot splicer. To keep track of all of these shots, the ribbons of cut film would be hung on a rail with a box below it (to catch the tail ends). This box was called a film bin. The process was slow and tedious, but very direct, and the editor felt in control. There's nothing quite like touching the actual frames of the shot. You could actually see how long a shot was, by looking at the film strip. The process was not

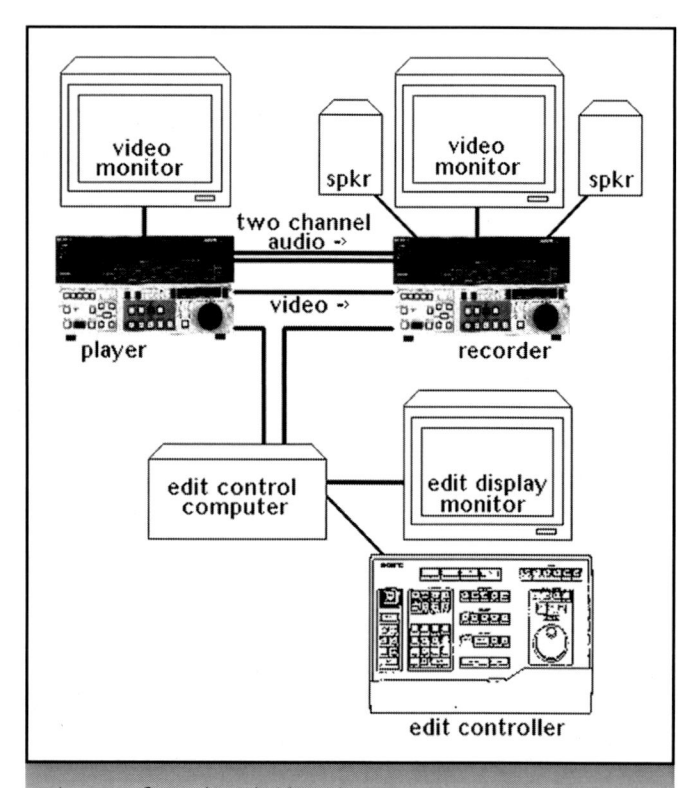

Diagram of tape-based video editing system

perfect, however. Film would get shredded or sometimes lost somewhere in the bottom of the bin (or on the cutting room floor).

For television people, when videotape came along, it was a whole new ball game. There was no developing of film negatives or work prints; the shots were simply reproduced on a second VTR. However, once you had laid down 50 shots or so, it was very difficult to insert another piece between, say, shots 27 and 28. Your choice was to insert the extra shot, and overwrite the rest of your hard-earned editing session, or make a dub of the end of the edit, insert the shot, and dub back the remaining program again (losing two generations in the process). More than just picture quality had been lost in the editing process, which can be so much a matter of cutting and recutting, inserting new material, and experimentation.

Nowadays, nonlinear editing systems digitize and store analog footage onto computer hard disk drives, providing random access from that digital storage. The editing process takes place on a computer, running appropriate software to perform various functions. The video and audio information is stored on large computer hard drives, where it can be viewed, modified, and eventually played back in real time from the system. The concept of having quick access to video information is a very powerful one. This allows editors to arrange and rearrange material to their heart's content, and then the work can be output directly to air or stored for later retrieval.

Today, these systems have on-line quality output and user-friendly interfaces. Nonlinear video editing is now cost effective for a wide range of applications including nonprofessional use as well.

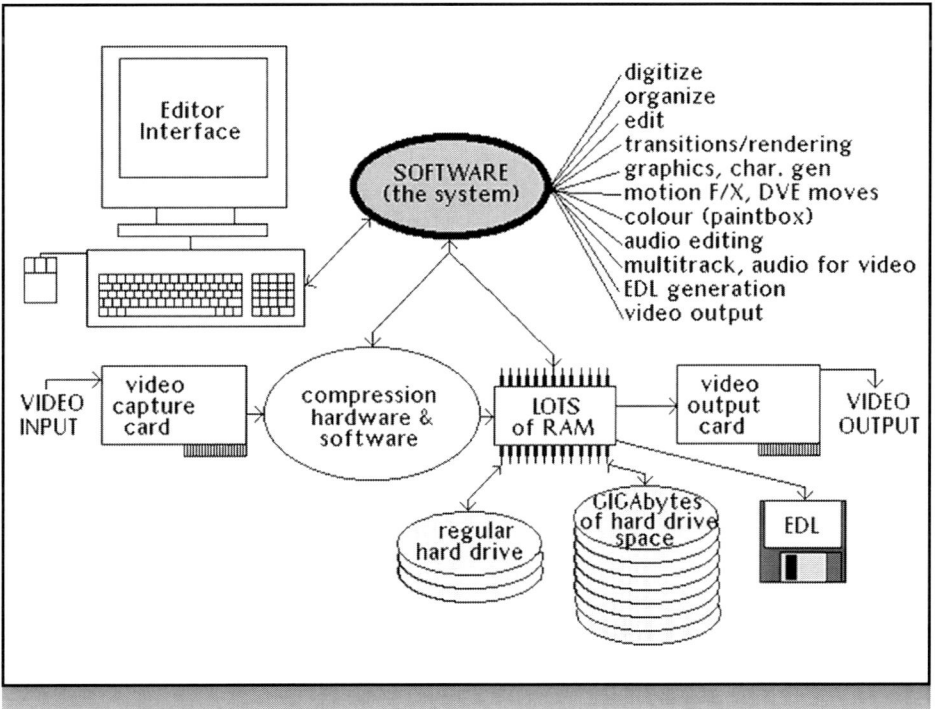

Nonlinear edit suite flow diagram

Typical nonlinear editing interface

In most systems, shots are displayed on the computer screen as being stored in a list, contained within a bin—the old film term comes back. This list can be displayed and sorted according to name, length, or any of several other categories. To view any shot, you just click on it with a mouse, and drag it onto the source screen where it's available for viewing. With nonlinear, there is no waiting for tapes to spool to preview a segment.

One of the best things about nonlinear editing is that the edit is instant: no splicing and waiting for the glue to set, as with film, and no having to actually play back the entire shot for its full duration, as with videotape. One mouse click and it's done—on to the next edit. And the shot can, of course, be placed anywhere, even in between two frames of a previously laid down shot.

14

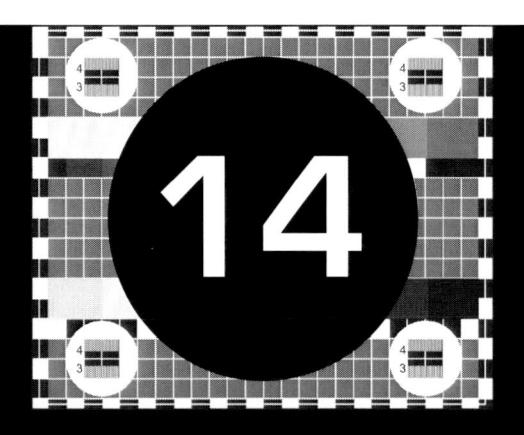

FILM FOR TELEVISION

"This film needs a certain something. Possibly burial."
—David Lardner, reviewing *Panama Hattie*, 1942

Overview

Film, as a finished edited and printed product, used to be one of the major program sources in television, but now it's pretty much restricted to movies and, occasionally, documentaries. Much program material is still shot on film, but it is immediately transferred to nonlinear editing systems for editing and distributed on either videotape or as a video file.

Film is classified according to its width (8mm, Super 8, 16mm, 35mm, 70mm and more) and its sound track type (silent, optical sound, magnetic sound). Television film projectors can accommodate either optical or magnetic sound.

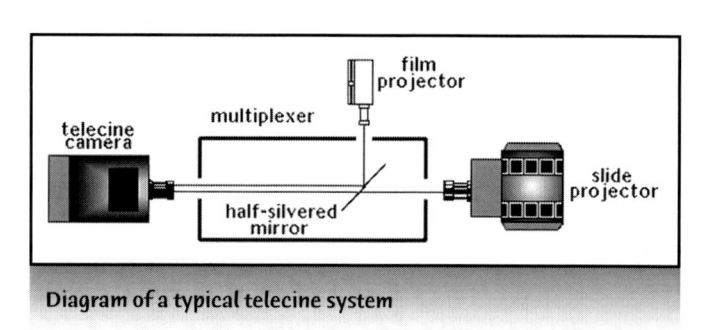

Diagram of a typical telecine system

Telecine

The first film transfer device used in television was called a telecine chain, also known as a film chain or film island. Telecines consisted of at least one film projector, a slide projector, a multiplexer (selector for film or slides) and a telecine camera.

Projector

The television film projector was specially designed so that film (generally 16mm), running at 24 frames per second, synchronized with the television rate of 30 frames per second. This was accomplished by scanning the first frame of film twice (two fields), the next frame three times (three fields), the third twice, and so on. This is called "three-two pulldown" (also written as 3:2 pulldown.)

How 3:2 Pulldown Works		
Original Film Frames	**Video *Fields***	**Resulting Video *Frames***
Film Frame 1 (shown for three video fields)	Video Frame 1, Field 1	> Video Frame 1
	Video Frame 1, Field 2	
Film Frame 2 (shown for two video fields)	Video Frame 2, Field 1	> Video Frame 2
	Video Frame 2, Field 2	
Film Frame 3 (shown for three video fields)	Video Frame 3, Field 1	> Video Frame 3
	Video Frame 3, Field 2	
	Video Frame 4, Field 1	> Video Frame 4
	Video Frame 4, Field 2	
Film Frame 4 (shown for two video fields)	Video Frame 5, Field 1	> Video Frame 5
	Video Frame 5, Field 2	

Film to video conversion, showing 3:2 pulldown process

There were some other important differences between regular and telecine film projectors:

- the ability to come up to full speed within a second or two, thus eliminating the need for a preroll
- easy, open-to-view threading mechanisms
- remote start and stop controls
- the ability to show a single frame for some time, for cueing purposes, without burning through the film
- projection and optical sound lamps that automatically exchanged themselves with a spare, should they fail during projection

Slide Projectors

Slide projectors had two vertically or horizontally arranged drums, usually holding about 18 slides each. Most were designed for forward and reverse action; some had a random selection, programmable system. While slides were easier to use than studio cards, the fixed telecine camera could not move over the frame or zoom in on particular areas of a slide. Also, vertically formatted 35mm slides could not be used in a telecine slide projector; the top and bottom of the frame would be cut off when projected through this system.

Multiplexer/Telecine Camera

The multiplexer was the series of mirrors or prisms that directed the images from the projectors into the telecine camera.

The telecine camera was similar to a regular television camera. Most had some form of electronic or electromechanical means of automatic brightness control. Some also had internal or external automatic colour correction systems.

Flying Spot Scanners

In the 1970s, another type of film transfer system was developed, called the flying spot scanner. It involved a CRT with a bright phosphor and a standard scanning pattern, focused on each frame of film. As the tiny spot of light passed different points on the film, it picked up the colour information for that particular frame. It then passed through a series of dichroic filters to separate it into its red, green, and blue components. Three photocells detected the amount of RGB information at any particular spot. These

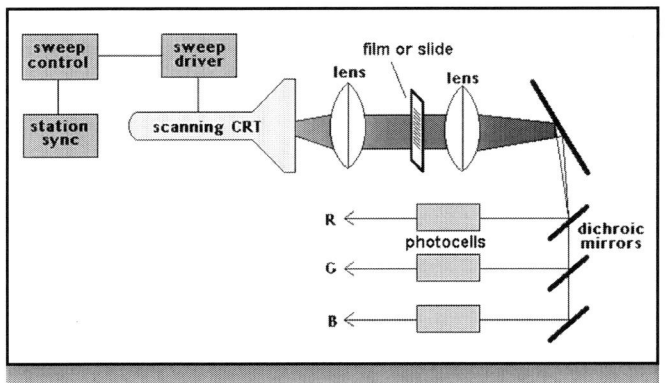

Principles of flying spot scanner telecine transfer

Flying spot scanner

three outputs were made available to colour correction devices and were then colour encoded in the usual fashion and presented to monitors, video recording devices, and the like.

The advantage of this system is that the detail and resolution of the scanned image was usually far superior to standard telecine transfer techniques. Also, because of the pattern of the scanning, the film could be transferred through the system in a more gentle fashion than the "pulldown claw" process used in standard telecine projectors.

Another, less obvious, advantage to flying spot scanning systems is that it was a simple matter to zoom in on film or a slide. All that was involved is that the scanning would take place over a smaller area of each frame. This allowed motion picture films to be scanned in any of several different formats. The more common ones are "letterbox" (where the full film is shown on the television screen, but there are black bands at the top and bottom of the 4:3 television monitor), and "pan and scan" (where the wider aspect film is panned by a 4:3 TV ratio area to show the most important action of a scene.) With the high information density of film and the relatively high resolution of the flying spot scanner, an image section can be magnified several times without graininess or loss of picture detail.

Flying spot scanners are still in popular use today, but the systems have been upgraded, and now use high-resolution CCDs as the image capturing device.

CONVERTING VIDEO TO FILM

Television shows (and especially, commercials and music videos) are often shot on film, as it imparts a certain look and feel to the project. The contrast ratio is much higher in film than in videotape, and there is a subtle grain that permeates the footage. However, film can be a rather expensive format on which to shoot. The original stock can't be reused, and there are costs involved with developing of the negative, work printing, editing, negative cutting, optical effects, release prints, and so forth.

There are now many methods readily available to convert the sharp look of video to the softer appearance of film. Finished video footage can be converted, frame by frame, through editing software that uses a plug-in to add grain, softness, and even scratches to your production.

15
TRANSMISSION

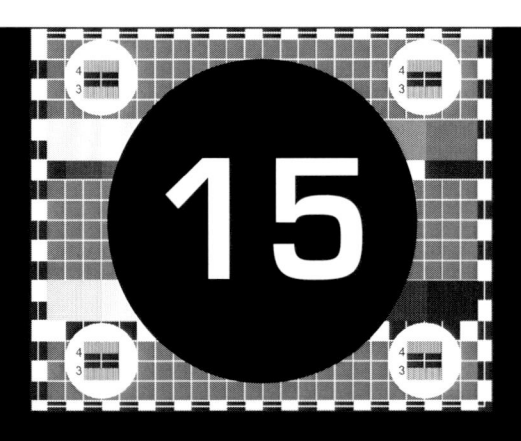

*"One night I walked home very late
and fell asleep in somebody's satellite dish.
My dreams were showing up on TVs all over the world."*

—Steven Wright

Basic Principles

Up until now, we've been discussing the principles of how to produce television sound and pictures. All of this is of little use if we have no way of actually sending this program material to our viewers. This chapter will deal with the challenge of transmitting our telecasts to the public at large.

Radio

We transmit our television programs using radio frequency (RF) waves. A look at how radio itself is sent will provide us with a solid background with which to study television broadcasting.

Radio relies on the radiation of electromagnetic energy from a transmitting antenna in the form of radio waves. These radio waves, travelling at the speed of light (300,000 km/sec, or 186,000 miles/sec), carry the information. When the waves arrive at a receiving antenna, a small electrical voltage is produced. After this voltage has been suitably amplified, the original information contained in the radio waves is retrieved and presented in an understandable form from a loudspeaker.

Transmission

At the heart of every transmitter is an oscillator. The oscillator produces an electrical signal of a given frequency, accurately controlled by a quartz crystal. After being amplified several thousand times, this voltage becomes the radio-frequency carrier. How this carrier is used depends upon the type of transmitter.

If applied directly to the antenna, the energy of the carrier is radiated in the form of radio waves. In early radiotelegraph communications, the transmitter was keyed on and off using a telegraph key or switch. The information to be sent was transmitted by short and long bursts of radio waves that represented letters of the alphabet by a sequence of dots and dashes known as Morse code, named after its inventor Samuel B. Morse. This type of transmission, known as continuous wave, is still used by amateur radio operators around the world. It can be found in modified form in high-speed teletype, facsimile, missile-guidance telemetry, and some space satellite communication. In these cases, the carrier is not switched off but shifted slightly in frequency; these shifts in frequency are decoded in the receiver and converted into computer data.

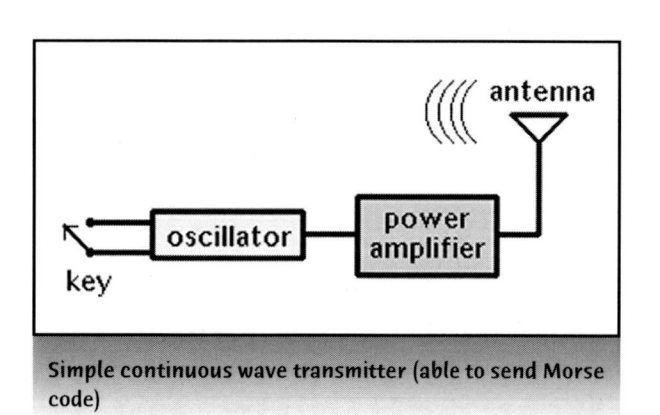

Simple continuous wave transmitter (able to send Morse code)

Amplitude Modulation (AM)

Simple AM radio transmitter

Frequency Modulation (FM)

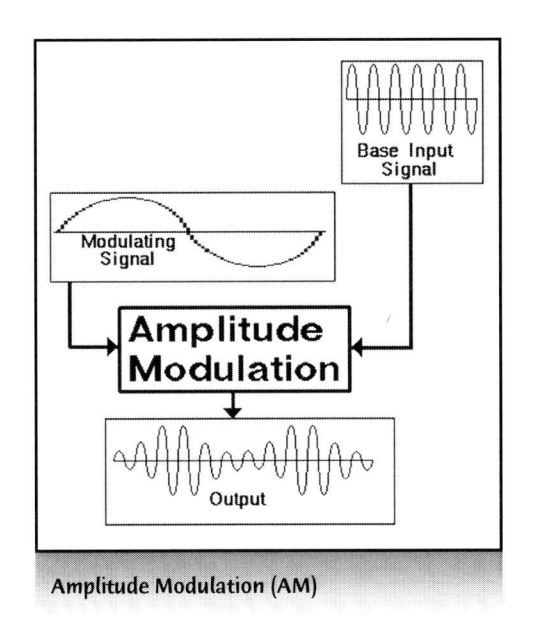

In standard broadcast transmissions, speech and music, instead of a switch or key, are used to modulate the carrier. One method is to superimpose the sound on the carrier by varying the amplitude of the carrier, hence the term amplitude modulation (AM). The modulating electrical signals that represent audio are amplified and applied to a modulator. When the audio signals go positive, they increase the amplitude of the carrier; when they go negative, they decrease the amplitude of the carrier. The amplitude of the carrier now has superimposed on it the variation of the original audio signal, with peaks and valleys dependent on the volume of the audio input. The carrier has been modulated and, after further amplification, is sent to the transmitting antenna.

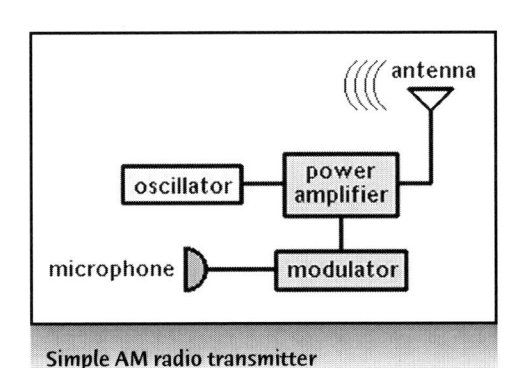

The maximum modulating frequency permitted by AM broadcast stations is 5 kHz at carrier frequencies between 520 and 1,710 kHz. The strongest AM stations have a power output of 50,000 watts, and can be heard sometimes for hundreds of kilometres.

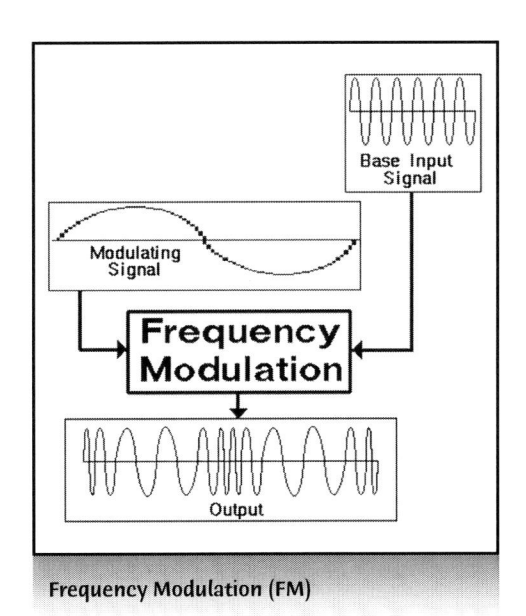

Another method of modulating the carrier is to vary its frequency. In frequency modulation (FM), during the positive portion of the audio signal, the frequency of the carrier gradually increases; during the negative period, the carrier frequency is decreased. The louder the sound being used for modulation, the greater will be the change in frequency. A maximum deviation of 75 kHz above and below the carrier frequency is permitted at maximum volume in FM broadcasts.

The rate at which the carrier frequency is varied is determined by the frequency of the audio signal. The maximum modulating frequency permitted by FM broadcast stations is 15 kHz at carrier frequencies between 88 and 108 MHz. This wider carrier frequency (15 kHz for FM as opposed to 5 kHz for standard AM broadcasts) accounts for the high fidelity of FM receivers. FM stations range in power from 100 watts to 100,000 watts. They cover distances of 25 to 100 kilometres, because FM relies on line-of-sight transmission.

Analog television transmitters use both AM and FM: the video, or picture, signals are transmitted by AM, and the sound by FM.

Antennas

An antenna is a wire or metal conductor used either to radiate energy from a transmitter or to pick up energy at a receiver. It is insulated from the ground and may be oriented vertically or horizontally; this is known as its polarity. An AM broadcast antenna is vertically polarized, usually requiring the receiving antenna to be located vertically as well. Television and FM broadcast transmitters traditionally have used a horizontal polarization antenna, although many FM and TV stations are now circularly (horizontally and vertically) polarized.

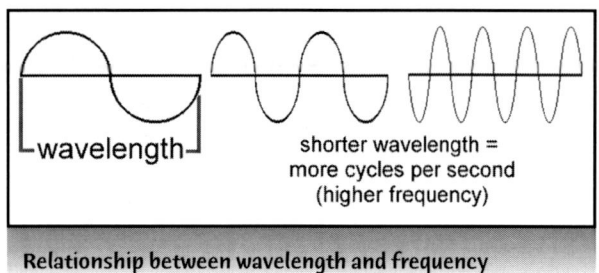

Relationship between wavelength and frequency

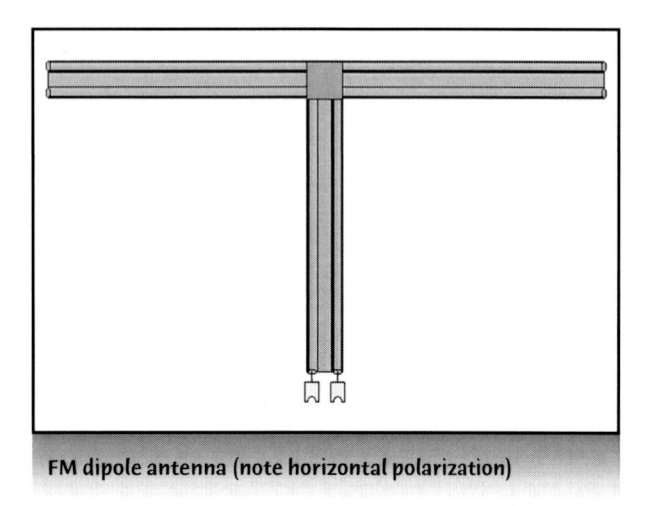

FM dipole antenna (note horizontal polarization)

AM transmitting field

Wavelength

There is a definite relationship between the frequency of a signal and what we call its wavelength. The higher a frequency is, the shorter its wavelength will be. Conversely, the lower the frequency, the longer the wavelength. There's an easy-to-remember formula for this. It's $\lambda = 300/f$, where λ is the wavelength you're trying to determine (in metres), and f is the frequency you're working with (in MHz).

For efficient radiation, the required length of a transmitting (and receiving) dipole antenna must be half a wavelength or some multiple of a half-wavelength. Therefore, an FM station that broadcasts at 100 MHz—which has a wavelength of 3 metres (300/100, using the formula above)—should have a horizontally polarized transmitting antenna 1.5 metres in length. Receiving antennas should be about the same length and placed horizontally; these sometimes take the form of rabbit ears for television antennas, or those small T-shaped wire FM dipole antennas that come with new FM receivers.

For an AM station broadcasting at 1,000 kHz (1 MHz), the half-wave transmitting antenna length should be 150 metres (300/1/2). As this is roughly the height of a 50-storey building, this is a bit impractical, especially when you consider it should be mounted vertically. In this case, a quarter-wavelength antenna is often used, with the ground (earth), serving as the other quarter wavelength. With these numbers in mind, it is easy to see why AM radio transmitters require large open spaces (often fields in less densely populated areas). FM transmitters can be easily located in more compact spaces.

Reception

When the transmitted carrier reaches the receiving antenna, a small voltage is induced into it. This may be as small as 0.1 microvolt (1/10th of a millionth of a volt), but is typically 50 microvolts for a standard AM broadcast. In the earliest form of AM radio reception, this voltage was coupled to a tunable circuit, which consisted of a coil of wire and a variable capacitor. The capacitor had a set of fixed metal plates and a set of movable plates. When the movable ones were adjusted, the capacitance was changed, making the circuit sensitive to a different, narrow frequency range. The listener selected, by adjusting the variable capacitor, which of the many transmitted signals picked up by the antenna the receiver would reproduce.

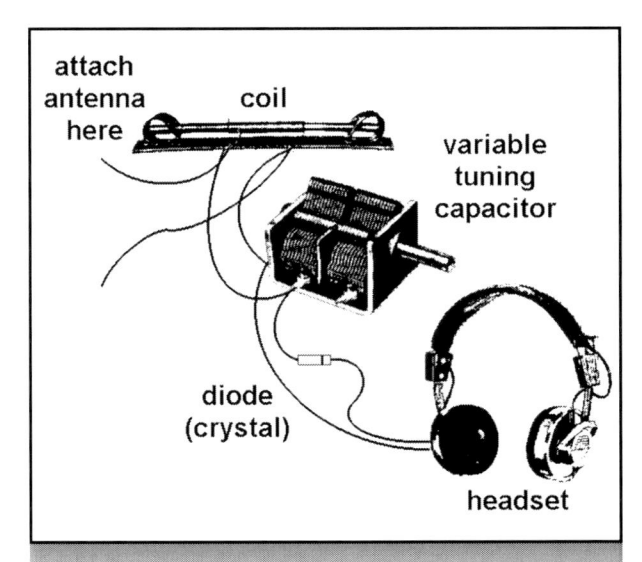

This early method of detecting radio waves was called a crystal receiver, or crystal set. A crystal of galena or carborundum with a "cat's whisker" provided a simple rectifier. The cat's whisker was a piece of fine wire delicately adjusted to rest upon the crystal in a sensitive place so that the rectification effect would take place. This stripped off the radio carrier wave, leaving only the audio information. Once detected, the audio was left to operate the earphones. Since no external electrical power or amplifiers were used, the only source of power in the earphones was the incoming signal. Only strong signals were audible, but with a long antenna and a good ground, reception of a signal from several hundred kilometres away was sometimes possible.

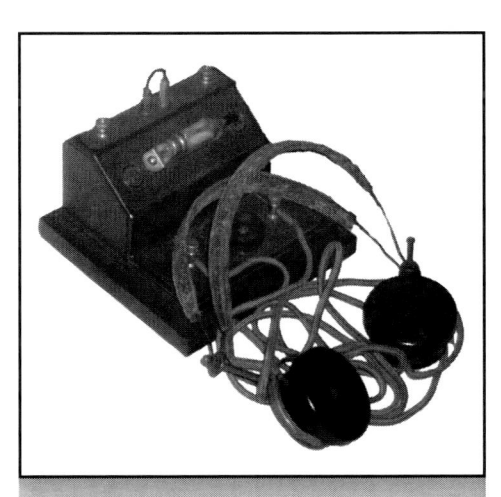

Following the development of the triode vacuum tube, increasing selectivity (ease of separating individual stations), sensitivity (how well distant stations can be received), and audio output power was possible. The tuned-radio-frequency (TRF) process involved several stages of radio-frequency amplification before the detection stage. In early receivers each of these stages had to be separately tuned to the incoming frequency, a difficult task at the best of times. Even after single-dial tuning was achieved by ganging together the stages, the TRF was susceptible to breaking into oscillation and was unsuitable for tuning over a wide range of frequencies. The principle is still used, however, in some modern shipboard emergency receivers and fixed-frequency microwave receivers.

Simple heterodyne receiver

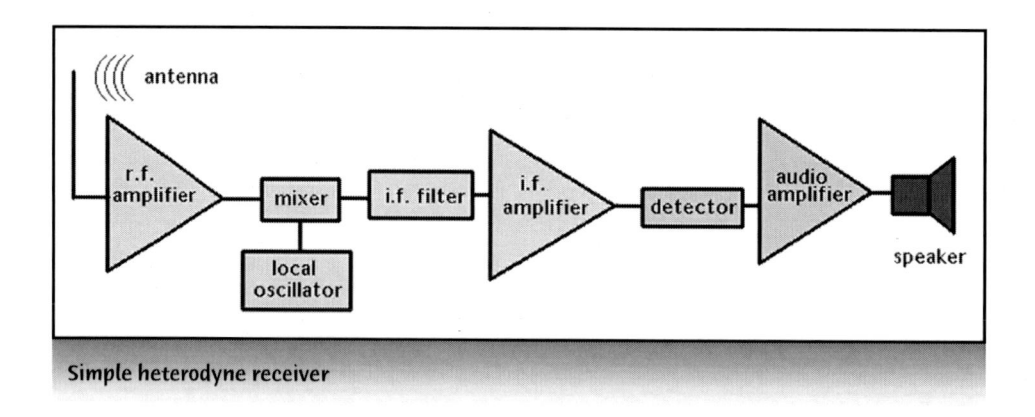

Practically all modern radio receivers use the heterodyne principle. All of the radio stations on a particular radio band (for example, AM radio) with their various modulated frequencies are amplified together, as a group. They are then combined with the output of a tunable oscillator inside the radio receiver, whose frequency is always a fixed amount above the incoming signal. This process, called frequency conversion or heterodyning, takes place in a mixer circuit. The output of the mixer contains the original carrier information of a particular tuned radio station, but converted to a single frequency. This is called the intermediate frequency (IF), and is typically 455 kHz in AM broadcast receivers, and 10.8 MHz in FM receivers. No matter what radio station the receiver is tuned to, the intermediate frequency is always the same. As a result, all further stages of radio-frequency amplification in the receiver can be designed to operate at this fixed intermediate frequency. This frequency is filtered and amplified, then detected, which means that the carrier is removed, leaving just the audio information. After detection, audio amplifiers boost the signal to a level capable of driving a loudspeaker.

Although the method of detection differs in AM and FM receivers, the same heterodyne principle is used in each. An FM receiver, however, usually includes automatic frequency control (AFC). If the frequency of the radio's oscillator drifts from its correct value, the station will fade. To avoid this problem, a voltage is developed at the detector based on the audio signal strength, which is then fed back to the local oscillator. This voltage is used to change the frequency output of the heterodyning oscillator to maintain the proper intermediate frequency and the best reception.

Both AM and FM receivers use automatic gain control (AGC), sometimes called automatic volume control (AVC). If a strong station is tuned in, the volume of the sound would be overwhelming if the volume control had previously been set for a weak station. This drawback is overcome by the use of negative feedback: a voltage is developed at the detector and used to reduce automatically the gain, or amplification, of the IF amplifiers.

The prime advantage of FM over AM, in addition to its fidelity, is its immunity to electrical noise, which imposes itself on an AM signal by increasing the amplitude of the signal. This effect shows up in AM as a crackling or buzzing noise. FM doesn't have this problem, because it decodes only the *frequency* variations, and has a limiter circuit that restricts any amplitude variations that may result from such interference.

Sidebands

When an audio signal of, say, 5 kHz is used to amplitude modulate a carrier, the output of the transmitter contains sideband frequencies in addition to the carrier frequency. The upper sideband frequencies extend to 5 kHz higher than the carrier, and the lower sideband frequencies extend to 5 kHz lower than the carrier. In normal AM broadcasts both sidebands are transmitted (and, by federal regulation, are limited to 5 kHz). This requires a total bandwidth in the frequency spectrum of 10 kHz, centred on the carrier frequency. It also accounts for why we perceive AM radio as being such low fidelity—it legally has a 5 kHz upper limit on the audio frequencies it can transmit.

unmodulated carrier generates a single frequency in the radio spectrum

modulated carrier generates its base frequency plus two sidebands

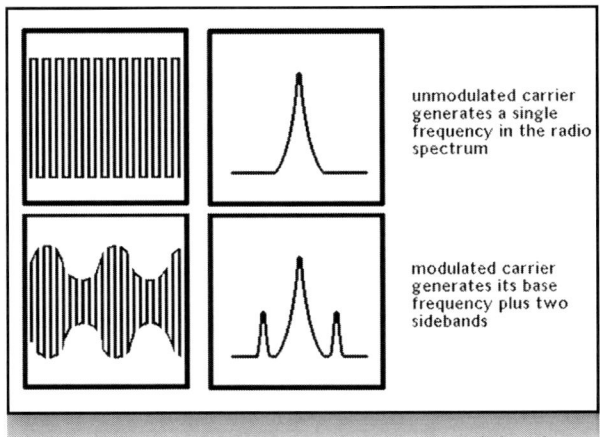

Sidebands generated when an AM carrier is modulated

The audio signal, however, is contained in (and may be retrieved from) either the upper or lower sideband. Furthermore, the carrier itself contains no useful information. Therefore, only one sideband really needs to be transmitted. A system designed to do this is called single sideband suppressed carrier (abbreviated SSB). This is an important system because it requires only half of the bandwidth needed for ordinary AM, thus allowing more channels to be assigned in any given portion of the frequency spectrum. Also, because of the reduced power requirements, a 110-watt SSB transmitter may have a range as great as that of a 1,000-watt conventional AM transmitter. Most ham radios, commercial radiotelephones, and marine-band radios, as well as citizens band (CB) radios, use SSB systems. Receivers for such systems are more complex, however. They must reinsert the nontransmitted carrier before successful heterodyning can take place.

The concept of sidebands will be important to us when we look at how an analog television channel is transmitted.

TELEVISION

Analog Transmission

When considering the transmission of standard definition television pictures, we must recall certain aspects of the video signal. The scene is scanned in about 1/30 of a second, and during that time about 280,000 picture elements must be covered. This number corresponds to scanning at the high rate of 4,200,000 Hz (with twice that number of picture elements) per second.

The transmission of the video signal at this fast a rate requires a wide channel in the radio spectrum. Each analog television channel in Canada occupies a frequency range of 6 MHz. This is 600 times as wide a band of frequencies as that

Television Technical Theory

Television transmitter

used by an AM radio broadcast station. The 6-MHz channel used is so wide that the radio spectrum has room for only 68 over-the-air channels. They are assigned among cities and towns at sufficient geographic and frequency separations so interference between channels does not occur.

Most of the television channel is used to transmit the video signal, which occupies a band of 5.45 MHz. A separate signal within the channel is used to broadcast the sound portion of a television station by FM. The high quality sound that frequency modulation can achieve is sometimes not heard in television receivers, because loudspeakers small enough to fit into portable televisions cannot reproduce bass notes properly. Fortunately, this is changing as manufacturers realize there is a market for high fidelity television audio.

The carrier signal for a television channel is an alternating current of very high frequency. On channel 2, for example, the picture carrier frequency is 55.25 MHz. This signal is generated initially by a quartz crystal oscillator at a lower frequency, which is multiplied and amplified until it reaches a power level of many kilowatts. The video signal controls one of the amplifiers, changing its power output. This amplitude modulated carrier is directed through the transmitting antenna.

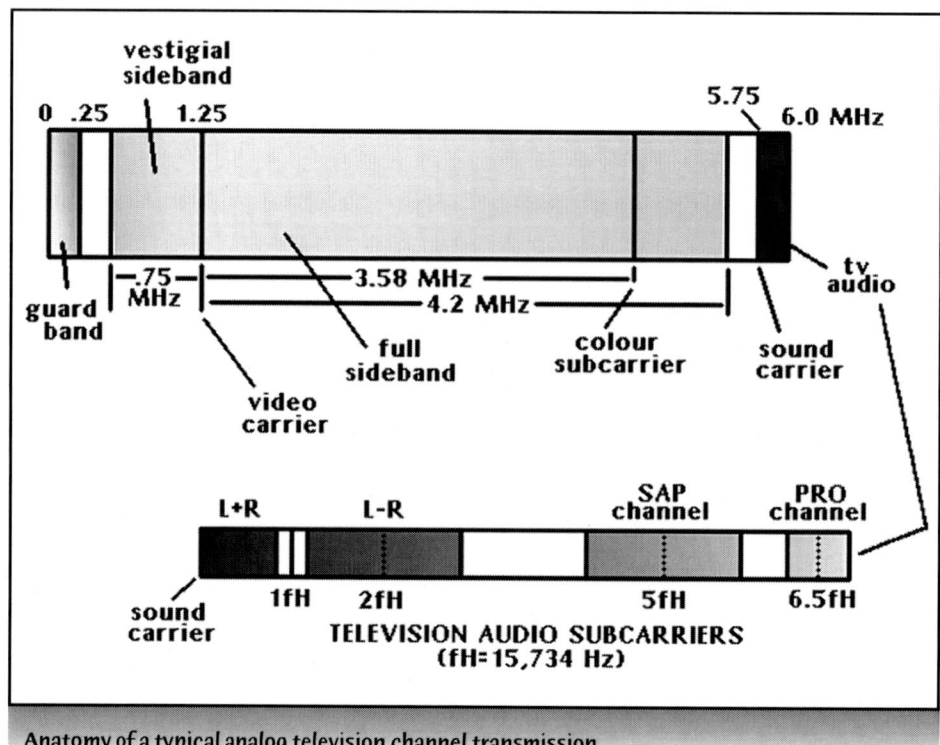

Anatomy of a typical analog television channel transmission

The amplitude of the radiated wave continually changes in response to the video signal it carries. More power is radiated during the dark portions of the picture, less power during the bright portions, with maximum power output during the synchronization pulses. This is the opposite of what you would first expect, considering that baseband analog video's highest output is at maximum white level, and the lowest level is at sync. The advantage of this so-called negative transmission is that noise pulses interfering with the transmitted signal increase the carrier amplitude toward black, which makes the noise less obvious in the picture. Also, the transmitter uses less power, with lower carrier amplitudes, for pictures that are mostly white or bright in intensity.

When the video carrier signal is modulated, two sidebands are produced. One of these, occupying a space of about 4 MHz, is transmitted in full, but only a portion, or vestige, of the other is radiated. This technique, called vestigial sideband transmission, saves a substantial amount of valuable space in the radio spectrum; otherwise, a channel almost 9 MHz in width would be required for each television signal.

Video signals transmitted in this way have the picture carrier at 1.25 MHz above the lower frequency boundary (this area being used by the smaller sideband), and extend up to 4 MHz beyond the carrier to 5.25 MHz. The colour subcarrier is placed at 3.58 MHz above the picture carrier (about 4.8 MHz within the channel). The sound carrier is placed .25 MHz (25 kHz) below the upper 6-MHz boundary; it is a conventional FM signal, with a bandwidth of about 50 kHz.

WHAT ARE ALL THOSE AUDIO SIGNALS?

Analog television audio's complexity has increased somewhat since the monaural days of the 1950s. It now has a series of subcarriers, piggybacked onto the main channel's audio, to allow us to receive such things as stereo, SAP, and PRO.

Stereo is sent to the television set by using two channels: one of them is Left+Right (monaural) and the other is Left-Right (the difference between the left and right sound information). This means that small, inexpensive sets can receive mono TV audio, while more expensive units can extract full stereo by using a simple matrix to decode the discrete channels. The original left channel sound information is decoded by adding the two signals together [(L+R) + (L–R) = Left] while the right channel of sound is created by subtracting the transmitted signals [(L+R) – (L–R) = – Right].

SAP is an acronym for Second Audio Program, and was originally intended to provide a way for TV stations to broadcast in two languages at once. Many stations use SAP for descriptive video services, also known as open captioning—a narration of action on the screen for the visually impaired. Other stations rebroadcast weather information, or use it as a promotional barker channel—a continuously running advertisement for upcoming programs.

PRO is the PROfessional use channel. This is for internal use by TV stations for such things as cueing reporters in the field, and is sometimes used for telemetry that relays information about a station's transmitter back to master control. With the use of two-way radios and cell phones for in-the-field communication, the PRO channel is now rarely used as an intercom cueing system.

Reception

The radio waves used in television broadcasting travel in straight lines, are intercepted when they strike any large object, and are weakened when they meet the horizon. To reach the largest possible number of viewers, therefore, the transmitting antenna must be located as high above the local terrain as possible. The primary service area of an analog television station thus seldom extends beyond 35 kilometres, although marginal reception is often possible at 75 kilometres if a highly sensitive receiving antenna is used.

Ghosts

When the television signal is intercepted by a nearby structure, such as a building, it is reflected in all directions. A receiver located between the transmitter and the reflecting structure, then, receives two signals, one directly from the transmitter as intended, and the other by reflection from the structure. The reflected signal, having travelled a greater distance, arrives later than the direct signal.

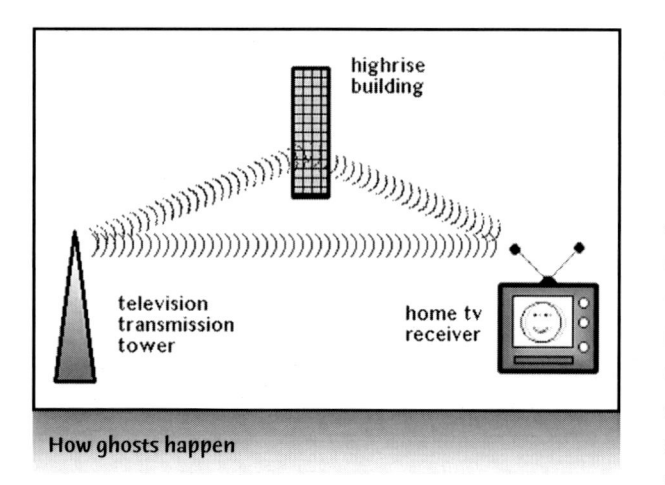

At the speed of light, radio waves cover about 300 metres in a millionth of a second. Hence, if the reflected path is longer than the direct path by 3 km, for example, the reflected signal arrives 10-millionths of a second later than the direct signal. The scanning of a line of analog television takes about 60-millionths of a second. So, during the scanning of each line, both the direct and the reflected signals produce images, the reflected signal producing a ghost of the intended image. In this example, the ghost image would be to the right of the intended image by about one-sixth of the width of the picture. Conditions in which a reflected signal exists, called multipath reception, are common in built-up city areas having many tall buildings.

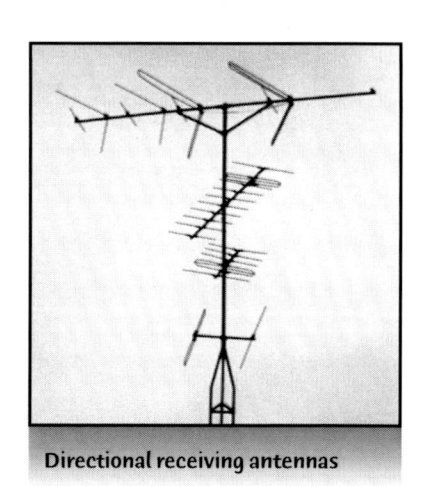

How ghosts happen

Directional receiving antennas

To lessen the effect, the receiving antenna must be as high as possible and oriented so it discriminates against the reflected signal. One method of avoiding reflections is to feed many home television receivers by coaxial cable, from a single community antenna located high above surrounding structures, where it is free from reflected signals.

The typical outdoor receiving antenna is constructed of several parallel horizontal metal rods of different lengths spaced one behind the other.

Such an array has directional properties, displaying maximum sensitivity on the line at right angles to the metal rods. For local reception, a less elaborate antenna will do, such as the extendable telescoping rod on a portable television receiver or the use of so-called rabbit ears.

Black-and-White Analog Television Sets

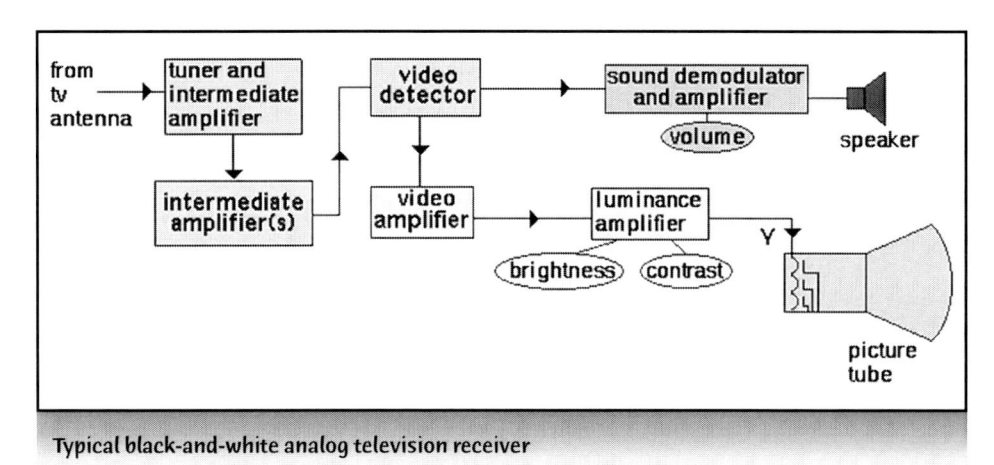

Typical black-and-white analog television receiver

In a typical black-and-white television receiver, the signal from the antenna is fed to the tuners. Two are used—one for the VHF (very high frequency) channels 2–13 and the other for the UHF (ultra high frequency) channels 14–69. Inside the tuner, the amplified signals from all TV channels in the area are passed to the mixer, which selects one channel, and transposes all the signal frequencies in the channel to different ones, called intermediate frequencies. The output of the tuner is fixed at an intermediate frequency range of 41 to 47 MHz, no matter what channel is tuned in.

From the tuner, the IF channel with all picture and sound information present is passed successively through several additional amplifiers (from two to four intermediate frequency amplifiers), which provide most of the amplification in the receiver. Their amplification is automatically adjusted, being maximum on a weak signal and less on a strong signal. So far, the receiver handles the signals in the channel just as they would be received from the transmitter, except for the shift to intermediate frequencies and the amplification.

The next stage is the video detector, which removes the high-frequency carrier signal and recovers the video signal. The detector also reproduces (at a lower frequency) the sound carrier and its frequency variations for FM audio. The sound signal is then separated from the picture signal and passes through a frequency detector, which recovers the audio signal. This signal is amplified further and fed to the loudspeaker, where it re-creates the sound. The picture signal from the video detector is used in the normal fashion for viewing on the video display device of the television receiver.

Colour Analog Television Sets

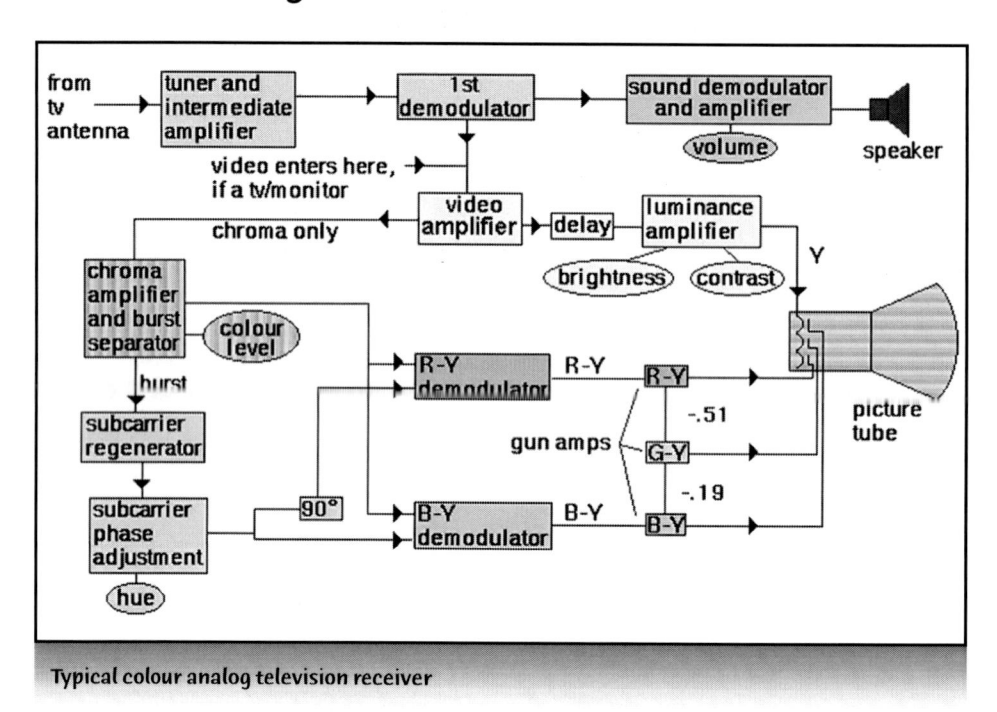

Typical colour analog television receiver

In a colour television receiver, additional circuits are provided to deal with the colour information.

The only difference in the IF circuit is the importance of bandwidth for colour receivers. Remember that video frequencies around 3.58 MHz just show up as fine details on a black-and-white TV set, but this frequency is essential for colour information decoding. Without it, there is no colour. This is why the fine-tuning control on analog colour television sets must be tuned exactly, or else the colour disappears, along with the higher resolution.

The sound is usually taken off before the video detector in colour sets, and a separate converter is used for it, instead of taking it from the video detector. The reason that this is done is to minimize a 920 kHz beat signal that can result between the 3.58 MHz colour subcarrier and the sound carrier signal. This signal would show up as interference in the television picture.

The output from the video detector is sent to two places: a series of colour circuits, and a luminance output amplifier. The luminance amplifier also serves as a cutoff filter for frequencies above 3.2 MHz, thus removing all colour information from the luminance video signal and, unfortunately, some of the sharpness and detail as well. This amplifier has the brightness and contrast controls for the TV set.

In the colour recovery circuits, several things happen. First, the video detector's output is sent through a colour band pass filter, which leaves us with just the chrominance information; the luminance has been removed. This chroma output contains both the colour information for the picture content and the colour burst. It is then sent to a burst separator to detect the phase and level of the colour burst. This is where we find the colour control on a TV set. At the burst separator, we now have a reference for the colours within one line of the picture content. This reference is sent to a crystal oscillator which generates a continuous 3.58 MHz subcarrier of the correct phase and amplitude based on the colour burst's levels and phase. This oscillator's phase can be adjusted, and this is the hue or tint control on a TV set. The oscillator is used with two colour demodulators to recover the R-Y and B-Y colour difference signals. The continuous wave subcarrier is delayed by 90 degrees of phase before it enters the R-Y demodulator. The R-Y and B-Y signals are each added to the luminance amplifier's Y signal to recover the red (R-Y + Y = R) and blue (B-Y + Y = B) colour channels. The green signal is created by subtracting the blue and red channel information from the Y signal (Y-R-B = G). The three colour signals are then sent to the picture display device.

DIGITAL TELEVISION TRANSMISSION

In the United States, over-the-air analog television transmission has now been replaced by digital television transmission. This will be happening soon in Canada as well.

The new ATSC (Advanced Television Systems Committee) standard, called A/53, involves using the same 6 MHz of bandwidth for each television channel, but changes the whole works over to a digital transmission system called 8VSB. The VSB stands for "vestigial sideband," which is not unlike the present analog transmission scheme in that sense. Digital TV sends over-the-air broadcasts by modulating the transmitter's carrier with digital information, instead of the traditional analog video signal. This improves the picture quality, versatility, and transmission efficiency, as well as adding the capability to transmit full 5.1 channel surround sound. One of the biggest advantages of the move to DTV is the ability to broadcast full high-definition video (HDTV) over an existing television channel. As well, DTV has eighteen different video formats that can be broadcast, from SD to HD, with frame rates from 24 to 60 frames per second, in both interlaced and progressive scanning methods. Because some of the lower resolution signals don't require the entire 19.39 Mbps digital channel, more than one of these can be sent simultaneously. These are called sub-channels. Keep in mind that the ATSC standard specifies details for transmission, not production or display.

Conversion of Digital Television Signals for Analog Television Sets

If a viewer is currently subscribing to a cable television or a DBS (direct broadcast satellite) service, their viewing is largely uninterrupted, as these systems continue to offer their set-top boxes which convert digital television back to analog for existing television sets. Even if a viewer currently uses a rooftop antenna or a small antenna in their living room to receive signals, their existing analog television set is not obsolete. These receivers can be augmented with a set-top box which converts over-the-air digital television transmissions into analog, so the sets can still be used.

Definition	Resolution (H x V)	Aspect Ratio	Frame Rate	Scanning Format
High Definition (HDTV)	1920 x 1080	16 x 9	24	progressive
			30	progressive
			60	interlaced
	1280 x 720	16 x 9	24	progressive
			30	progressive
			60	progressive
Standard Definition (SDTV)	704 x 480	16 x 9	24	progressive
			30	progressive
			60	progressive
			30	interlaced
	704 x 480	4 x 3	24	progressive
			30	progressive
			60	progressive
			30	interlaced
	640 x 480	4 x 3	24	progressive
			30	progressive
			60	progressive
			30	interlaced

DTV video formats

To create a digital television, over-the-air signal, several steps are undertaken. First, the video and audio from a program are converted to MPEG-2 (if they're not in that format already). The important thing about this particular MPEG stream, though, is that it is a constant 19.39 Mbps. This is called the DTV transport layer.

The MPEG stream is sent to the transmitter, where framing information is added. During this time, the signal is also randomized so that it is of a more or less constant signal level. The stream is then broken up into 207 byte packets. These packets are further broken up into four 2-bit words with error correction. Further sync signals are added to these small words: segment sync, field sync, and what's called the ATSC pilot, which is a signal that allows a digital television set to lock onto a particular channel.

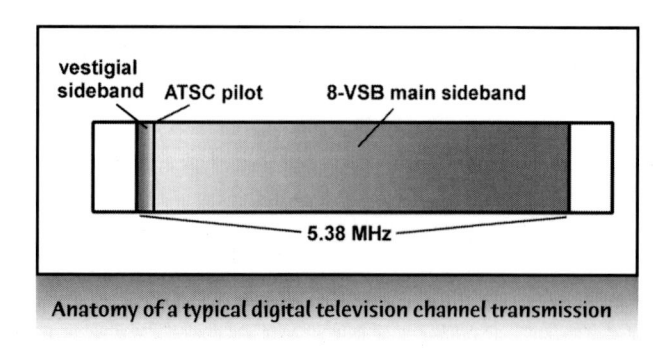

Anatomy of a typical digital television channel transmission
(vestigial sideband, ATSC pilot, 8-VSB main sideband, 5.38 MHz)

The final signal is sent to the transmitter's modulator, and is sent out into the airwaves on the proper TV channel. The signal still takes up almost the entire 6 MHz television channel (actually, closer to 5.38 MHz), but is of a very different format from an analog channel.

One more thing to keep in mind. Digital transmission, as we know, requires a certain minimum signal strength. If there is sufficient signal strength for the set to decode digital data, a perfect picture will result. If the signal strength falls below that level, the reception stops completely. There is no gradual deterioration of the picture quality as there is with the analog NTSC system. It is either all or nothing. This shouldn't be a problem for viewers within the normally accepted coverage area of a television transmitter, but people who previously received stations in fringe areas, with snowy but viewable pictures, may find that they are no longer able to receive their favourite long-distance programming.

The Electromagnetic Spectrum

Electromagnetic radio waves use the radio spectrum. The lowest frequencies have the longest radio waves; the highest frequencies, the shortest waves. The spectrum is split up into bands, since each portion of the spectrum has uses which are more appropriate than others, due to the physics of electromagnetic radiation and electronic design. All frequencies have been assigned by international agreement at a World Radiocommunication Conference (WRC). These are organized by the International Telecommunication Union (ITU), an agency of the United Nations.

After a WRC, or whenever Canada's needs change, Industry Canada allocates frequencies to services within the electromagnetic spectrum. A more detailed presentation of these allocations, including footnotes, can be found in Canada's Table of Frequency Allocations. It can be found online at Industry Canada's website, *http://www.ic.gc.ca* (search for "Radio Spectrum Allocations in Canada"). The radio spectrum is used by anyone who owns or operates any kind of a radio transmitter.

Where Does the Analog Cable TV Channel System Fit into This?

Cable TV had its start in small, isolated communities that couldn't receive television very well. A handful of people decided to buy an expensive antenna, and mount it in a good spot to receive signals. To offset the cost of the system, they decided to rent the strong signals that they were receiving to some other people. That's how cable television started, as a community antenna system (CATV).

Today, the round coaxial cable that comes into your home from the cable company does the same job as the old fashioned flat twinlead did in the days when everybody had their own rooftop antennas—it carries the signals to your set. Twinlead has a problem, though: long lengths of it act like an antenna, picking up all sorts of signals (that's why they still use it to make the FM antenna that comes with your stereo). So, for large systems with miles of cable, coaxial cable is used. The outside shield conductor keeps the cable company's signals in the cable, and hopefully keeps unwanted signals from the airwaves out of the cable.

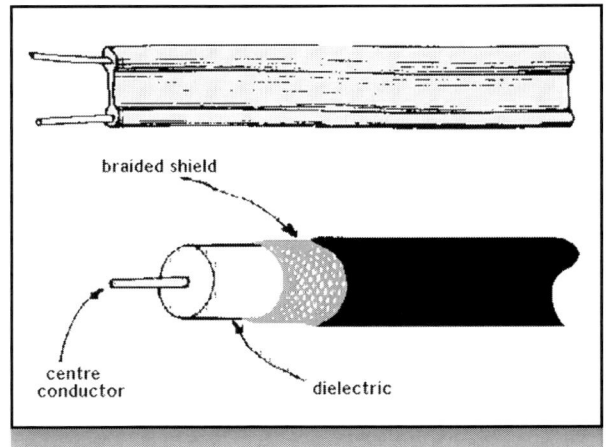

Twinlead (top) and coaxial cable (bottom)

Abridged Canadian Radio Frequency Spectrum Chart, With An Emphasis On Frequencies Used For Purposes Within The Broadcasting Industry

Frequency	Use	Cable TV Use
3-9 KHz	not allocated	
9-14 KHz	long range navigation	
14-535 KHz	an assortment of fixed and maritime mobile radio, radio location, radio navigation, maritime radio navigation, aeronautical mobile, aeronautical radio navigation and mobile radio	
535 KHz - 1.7 MHz	AM broadcasting	
1.7-54 MHz	various amateur bands, shortwave broadcasting, time signals, industrial, scientific and medical usage, CB radio, cordless phones and mobile radio	
54-72 MHz	TV channels VHF 2 through 4	Ch. 2-4
72-76 MHz	Aeronautical navigation and astronomy	
76-88 MHz	TV channels VHF 5, 6	Ch. 5-6
88-108 MHz	FM broadcasting (100 channels spaced 200 kHz apart)	Ch. 95-99
108-135 MHz	aeronautical navigation and communication	120-174 MHz: Ch. 14-22
135-144 MHz	space research, land mobile radio	
144-148 MHz	amateur radio	
148-174 MHz	land mobile radio, marine vessel traffic, government fixed mobile, paging services	
174-216 MHz	TV channels VHF 7 through 13	Ch. 7-13
216-470 MHz	an assortment of land mobile, amateur radio, Family Radio Service, government fixed and mobile radio, aeronautical, and marine radio	216-648 MHz: Ch. 23-94
470-608 MHz	TV channels UHF 14 through 36	
608-614 MHz	space operations (there is no UHF 37)	
614-806 MHz	TV channels UHF 38 through 69	
806-890 MHz	cellular telephones, trunked mobile radio	
890 MHz - 1.2 GHz	an assortment of radio location, land mobile and paging, amateur radio, aeronautical satellite, space observation, air traffic control radar, cordless phones, paging, maritime radio	
1.2 GHz	global positioning system (GPS)	
1.2-1.47 GHz	an assortment of radio location, amateur radio, aeronautical satellite, space observation, air traffic control radar	
1.47 GHz	digital audio broadcasting (DAB)	
1.535-1.56 GHz	mobile satellite (MSAT)	
1.575 GHz	global positioning system (GPS)	
1.626-1.66 GHz	mobile satellite (MSAT)	
1.9 GHz	personal communicatino service (PCS)	
2.475 GHz	digital cordless phones, wireless LAN, microwave ovens	
3.5-4.8 GHz	C band (Anik B, D, E) satellite downlinks	
4.8-5.8 GHz	various radio navigation	
5.8-7 GHz	C band satellite uplinks	
7-8.5 GHz	various uplinks and downlinks	
8.5-10.7 GHz	various radio navigation and location, including police radar	
10.7-12.2 GHz	Ku band (Anik B, C, E) satellite downlinks	
12.2-12.7 GHz	direct broadcast satellites (DBS)	
12.7-14.5 GHz	Ku band satellite uplinks	
17.5 GHz	direct broadcast satellites (DBS)	
20 GHz	multimedia satellite	
28 GHz	local multipoint communications systems (LMCS)	
29.75 GHz	multimedia satellite	
30-275 GHz	various extremely high frequency (EHF) allocations, many of them satellite usage	
275-400 GHz	not allocated	

Chapter 15   Transmission   215

The electromagnetic spectrum

3 kHz VLF
- 6 kHz — 14 kHz: Omega long range navigation | mobile radio, radio location and navigation

30 kHz LF
- 90 kHz — 110 kHz: Loran C
- mobile radio, radio location and navigation

300 kHz MF
- 535 kHz — 1.705 MHz: AM radio
- mobile radio, radio location and navigation

3 MHz HF
- amateur radio, shortwave broadcasting, industrial scientific and medical
- 27 MHz: C, B

30 MHz VHF
- baby monitors, old cordless
- industrial scientific and medical
- 54 — 72 76 aero — TV 2, 3, 4 — 88 FM radio — 108 117 aero nav — 137 aero control — space, ham, paging, mobile — 174 TV 7-13 — 216
- mobile, ham, aero, marine

300 MHz UHF
- 470 — 608-614 — TV 14-36 — TV 38-69 — 806 analog cell phones — 890 cordless — wireless mics — Cable 14-22 wireless mics — Cable 23-37
- Cable 2, 3, 4 — Cable 5, 6 — Cable 7-13
- GPS — DAB — GPS — MSAT — PCS phone

3 GHz SHF
- Cable 38-94
- 3.5 — 4.2 — 5.8 — 7.0 — 10.5 — 10.7 12.2 — 12.7 — 14.5 — 2.45 GHz
- C band down | radio navigation | C band up | up and downlinks, radio nav & beacons | Ku down D B S | Ku up
- police radar
- micro-ovens, digital cordless, wireless LAN

30 GHz EHF
- various EHF allocations, (including satellites), and unallocated frequencies

As long as the cable company has a cable from them to you, there are a lot of other signals they can send. Most come from satellite channels especially for cable viewers. But the regular UHF channels from 14 to 69 can't be sent down cable TV unmodified, because their high-frequency energy is absorbed by the cable itself. So, that leaves us with 12 channels (what would normally be on the VHF dial) to accommodate dozens of signals.

If you look at the government assignments for who can broadcast on what frequency, there's a gap between 6 and 7. There's also a huge gap between 13 and 14. The signals that the cable company sends are transmitted on the same frequencies as the ones that are sent by other types of transmissions, except the cable company's signals are confined to the coaxial cable. This means they can use the same frequencies that are occupied by aircraft, police, and taxis, and they won't interfere with those other services.

To pick up the signals resulting from this technology you'll need either a cable converter or a cable-ready television set. TV sets used to be built to tune in only VHF and UHF channels. While all sets are cable-ready now, they can still be set to receive the over-the-air spectrum. With all that converting of frequencies going on, the cable company gives you a chart to tune in your favourite over-the-air channels on their new cable channel allocations.

Here's what's happening. The numbers 2 to 13 on your converter are real channels 2 through 13. The cable company probably has converted some UHF channels in your area to these, and chances are that the others are on the wrong channels, but they really do occupy the standard channels 2 through 13. Channels 14 and up aren't what they seem—14 to 22 are the hidden 9 channels between the over-the-air channels 6 and 7, and 23 to 64 are bunched between the over-the-air channels 13 and 14.

For those of you with 125 channel cable-ready televisions, you also have channels 65 through 94, located in UHF areas 14 through 43. But this goes against what we saw earlier, when it was noted that high frequency UHF channels can't be transmitted down the cable. As it turns out, they can, but they become increasingly noisy as you go up in channel number. This is why, until lately, cable companies weren't using these frequencies. To finish up the cable TV conversion chart, cable channels 95 through 99 are located to cover completely, and go just above, the FM radio band.

DON'T TOUCH THAT DIAL

Digital Cable, Telcos, MMDS, LMCS, DBS

Over-the-air terrestrial transmission and standard analog cable TV no longer have a monopoly on the way to get television signals into your home. There are several other systems.

Digital Cable TV

Brought to you by your local cable company, it's a digital multichannel version of what we know in the analog world, using the same cable, but getting around that coaxial cable frequency-absorption problem by encoding the television channels digitally. It requires a set-top box to decode the digital signals back into analog NTSC for you to enjoy. Unfortunately, this new digital cable television system, known as QAM (quadrature amplitude modulation), is incompatible with the over-the-air DTV standard A/53.

Video by Telephone

Your local telephone company, for years, has been working to perfect a system that will bring video to your place via the phone lines using a system known as Internet Protocol Television (IPTV.) This is not to be confused with watching video over the Internet. IPTV uses digital network protocols to send a television channel to your home, but it runs over dedicated telephone lines, not the Internet. Some systems are in place in North America, but it hasn't gained the popularity of cable or satellite distribution systems.

Multichannel Multipoint Distribution Systems (MMDS)

MMDS works in the 2.5 to 2.686 GHz range and can do a line-of-sight transmission of 40 to 50 kilometres. The main antenna is usually on a high point in the city, but there are some booster/repeater towers around town, so if you can see any of those, you can also subscribe to MMDS distribution. The repeaters increase the coverage by receiving the main signal and retransmitting it further afield. MMDS uses a 30-cm-square flat antenna. The MMDS system is comparably priced to regular cable TV subscription rates, after an installation fee.

Direct Broadcast Satellite (DBS)

This system is direct-to-home satellite reception, using various pizza-sized dish antennas seen on the fronts of houses, and on rooftops and balconies.

The launch of Hughes Galaxy 601 (commonly known in the industry as DBS-1) in December 1993 signalled the beginning of a new era in entertainment distribution. This technology now beams hundreds of channels of audio and video programming in both standard and high definition, to 18-inch satellite dishes installed in homes all across Canada. The commonly used DBS satellite compression system is known as QPSK (quadrature phase shift keying) and it, too, is incompatible with the new over-the-air DTV A/53 system.

Wireless Mics, IFBs, Headsets, and Interference

Wireless audio and video gear allows us to shoot in any area, at any time. Wireless microphones, wireless IFBs, and intercom sets allow freedom of movement for various people involved in the broadcast or EFP shoot. Wireless video transmission equipment allows us to shoot on convention floors and other locations where stringing cable would be prohibitive.

Once thought of as exotic and temperamental, wireless mics have become tame over the past couple of decades. The increase in mobile cameras and recorders has ensured a solid place for them in the future of broadcasting. This equipment is very handy, but causes some problems since it can cause interference—and be interfered with. One should always have a backup (wired) system, in case of unforeseen difficulties.

Especially with VHF body-pack transmitters, it's a good idea to make sure the antenna on the talent is vertical. If the receivers' antennas are vertical, the transmitters' antennas should also be vertical. Talent may have to be discouraged from coiling up a VHF antenna and tucking it neatly away. However, a flexible wire whip antenna is still preferred to a more rigid rubberized antenna, since the human body is largely composed of water and salt (which means it conducts electricity), and it tends to detune the stiffer type of antenna.

To help eliminate the effects of RF dropout as the talent moves around the shooting environment, diversity reception has been developed. Today, most systems use two antennas with two receivers. A switch, sensitive to the levels of radio frequency from each of the receivers, selects the output from the stronger signal while ignoring the weaker. Some systems mix the two receivers, instead of switching from one to the other. Another system involves checking the received RF phase between two antennas. The phase of the second antenna is constantly adjusted to complement the signal from the first.

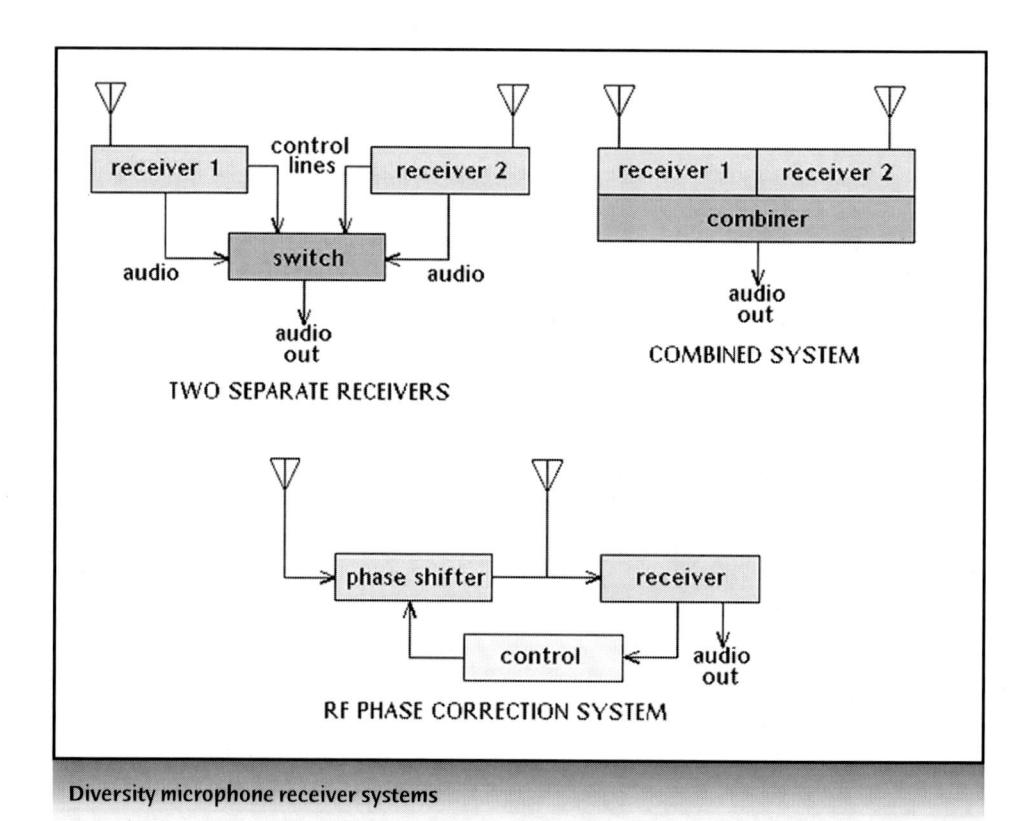

Diversity microphone receiver systems

Wireless audio gear operates on various VHF and UHF frequencies between television channels 7 and 13. Some stations have taken wireless equipment usage to a new high. As of this writing, there are transmitters on various frequencies between 174–216 MHz (interleaved neatly between and through channels 7 through 13); 470–608 MHz (interleaved between and through channels 14 through 36); 614–806 MHz (interleaved between and through channels 38 through 69) and 902–916 MHz (a new band for microphones). These frequencies are used for wireless microphones, IFBs, and floor director headsets.

Wireless video transmission gear operates in the 2 GHz range and include microwave trucks and short-distance wireless links, freeing up the cameraperson from being tethered to a video cable back to the remote truck.

SATELLITES

An artificial satellite, in general terms, is an object placed into orbit around the Earth for scientific research, radio or television transmission, or military reconnaissance.

How They Stay Up There

A theoretical object, orbiting just above the earth's atmosphere will neither crash into the ground, nor fly off into space when it's given a horizontal velocity of approximately 28,800 km/h. This is because, at this velocity, the Earth's surface curves away from the object as fast as gravity pulls it downward. This theoretical object would circle the globe in about 90 minutes. As the altitude of the satellite increases, its velocity can decrease. Its period, however—the time the satellite takes to circle the Earth—increases. These theories were first proposed by Arthur C. Clarke in the October 1945 issue of Wireless World.

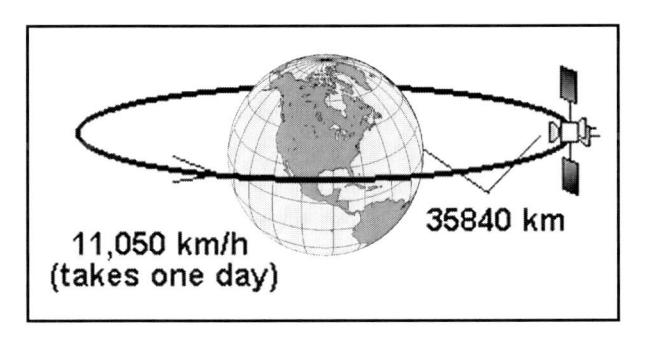

Based on these facts, if we were to take our satellite and move it up to 35,840 km above the Earth, it would have a period of 24 hours (the same amount of time as the Earth's rotation). It would have a velocity of approximately 11,000 km/h. A satellite like this is called a geosynchronous satellite. If such a satellite orbits above the equator, it is termed geostationary because it will remain at the same relative point above the Earth's surface. There are currently about 300 geostationary satellites orbiting above the equator, all around the world.

What's In a Satellite?

All artificial satellites have certain features in common:

■ radar for altitude measurements

■ sensors such as optical devices in observation satellites

- receivers and transmitters in communications satellites
- stable radio signal sources in navigation satellites
- antennas to receive and transmit signals
- solar cells to generate power from the sun, and storage batteries used for the periods when the satellite is blocked from the sun by the Earth. These batteries in turn are recharged by the solar cells
- in special cases, nuclear power sources are utilized
- attitude control equipment is needed to keep the satellite in its desired orbit and to point the antennas or sensors properly towards Earth
- telemetry encoders which measure voltages, currents, temperatures, and other parameters describing the health of the equipment, and relay this information back to Earth stations

Why Do We Use Them?

Most long-distance radio communication across land is sent via microwave relay towers. These towers, 30 to 60 metres high, are typically spaced 30 to 50 km apart, and about 100 of them are needed to cross Canada. Therefore, microwave relay towers are impractical for transoceanic communications.

The satellite serves as a sort of tall microwave tower to permit direct transmission between stations. But, unlike a microwave link or a cable, it can interconnect any number of stations that are included within the antenna beams of the satellite rather than simply the two ends of the microwave link.

Typical microwave tower

WORLD ANALOG TELEVISION STANDARDS

The three acronyms NTSC (National Television Systems Committee), PAL (Phase Alternate Line), and SECAM (Sequentiel Couleur Avec Memoire [sequential colour with memory]) represent the types of colour television systems used in the world. Such technical factors as number of video lines, field rate, video bandwidth, modulation technique and sound carrier frequency make up the differences within the three main methods. There are also several variations within each process.

If you understand the NTSC system, you have the basis for understanding PAL and SECAM. All three systems, for example, use the RGB principle of picking up the colour picture information from a scene. They also all include the idea of being compatible with previously invented monochrome standards. Therefore, the luminance information is chosen in all approaches to occupy the wideband portion of the channel and to convey the brightness as well as the detail information. Also, the chrominance information, in all systems, is superimposed upon this luminance signal. In many parts of the world, these systems are now being replaced with digital video and transmission systems. If the reader is interested in more details about international analog television systems, the Internet has a wealth of information on this topic.

Master Control

When all of the diverse elements of the daily program schedule come together, there has to be some way of integrating them in a connected way. That is where master control comes in. Master control operators (and their equipment) are responsible for the final output and look of the station, and carry out a variety of tasks:

- rolling in various prerecorded programming, either on videotape or from video servers

- taking live shows (such as news programming and flashes) at the proper time

- inserting at the proper time appropriate bumpers and commercial breaks, logging those commercials

- adding station IDs and audio voiceovers when required

- ensuring that all programming is integrated into the day in accordance with the directions of the programming and traffic departments

- making sure that the transmission of the station is always flawless

It is a sometimes hectic series of proceedings, involving minute-by-minute scrutiny of programs, cue sheets, and technical quality of all material. Logs of all commercial and promotional content have to be accurate, as required by federal law. There are also video logs to be maintained, which are kept in a library or on a server, containing every minute of broadcasting.

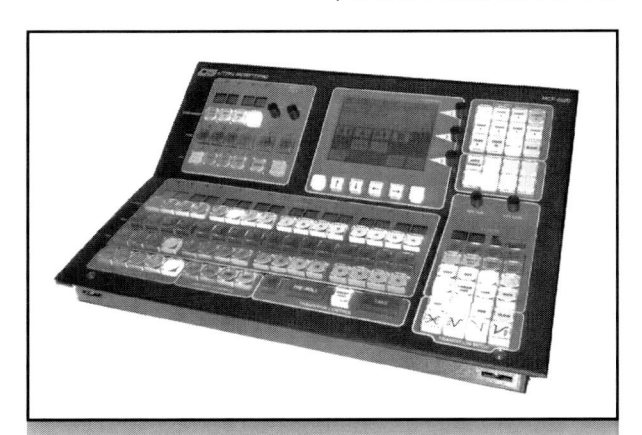

Master Control Switcher

The heart of the master control operator's equipment is the switcher, which can perform cuts, dissolves, and keys like a production switcher. There is one significant difference, however. The MCR switcher also has the ability to take audio from any of the sources selected, and is therefore called an audio follow switcher. In addition to the regular program audio, the master control room operator (or MCO) has the capability of sending out additional audio material, which can either replace the existing audio completely or be mixed over the continuing source. This is done by lowering the level of the program material and is called

Master control switcher

voice-over mode. The keyer on the switcher is generally used to superimpose station identification information over program material, or other statistics such as the time of day.

With all of this information and technical detail to watch over, master control operations are now mostly computerized. The computer remembers and activates transition sequences. It also cues, rolls, and stops videotape machines and servers and calls up any number of slides from a stillstore system.

16

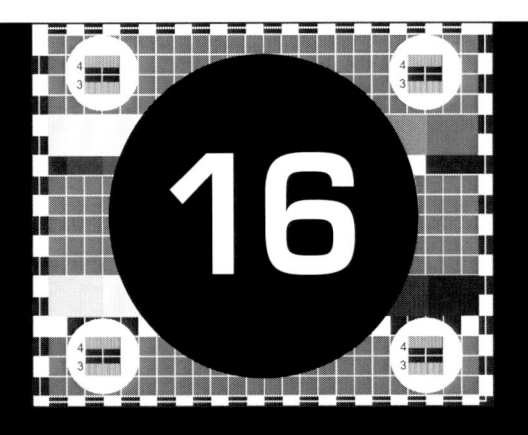

SECTION REFLECTIONS

WHAT'S IN A TELEVISION STATION?

CHAPTER 1

250 words

Television stations are rather holistic places. There are a lot of components, but they all fit together to make the television production process a smooth one. Try not to think about the system as a big lump of equipment, but rather consider the components used and their individual importance to the whole.

However, given your introduction to the complexity of a television station, have a look at all of the technology and operational positions and pick one that particularly interests you. Indicate which one(s) interest you the most, and state why.

TELEVISON: A BRIEF HISTORY AND OVERVIEW

CHAPTER 2

250 words

Answer one of the following questions:

1. TV started out with a mechanical system. What was it? In simple terms, how did it work?

2. Who are some of the key names in the invention and evolution of electronic television?

3. Who are the organizing bodies that gave us the television standards we use today?

4. In general terms, what was going on in Canada? Why? What is currently happening in the Canadian landscape, in terms of technological innovation?

CHAPTER 3 — ELECTRICITY 101

250 words

Answer one of the following questions:

1. Most circuits convert electricity into another form of energy. They have four main components. What are they? Why do you care?

2. There are series and parallel circuits. Compare and contrast the two types.

3. Why is this "electricity 101" information important to us as TV people?

4. Parallel and serial computer transmission differ. State the advantages and disadvantages to each, and give examples of their use.

CHAPTER 5: ANALOG VIDEO

250 words

The purpose of our legacy analog television system is to allow us to send television down a single transmission channel. Some parts of the process to do this include scanning, interlace, synchronization signals, and, in the case of colour television, colour encoding of separate colour channels of picture information.

This system has inherent within it certain limits. What are those limits? Why does this matter?

CHAPTER 6
DIGITAL VIDEO

250 words

Answer one of the following questions:

1. In the analog world, time is continuously observed. In the digital world, time is sampled. To convert our analog world to digital and back, we use analog-to-digital converters, and digital-to-analog converters. Converting our analog video to digital has several advantages and some disadvantages. State what these are, and give examples of their use.

2. Explain in your own words how MPEG compression works. Include advantages and disadvantages of using video compression in television production.

CHAPTER 7
MEASURING VIDEO

250 words

We need a way to view our video information other than monitors, so that we may maintain quality control over our television signal. Waveform monitors can display luminance and chrominance information, showing it to us in various ways and in various detail. Vectorscopes can display chrominance information—hue (colour phase) and saturation, again, in various ways.

Answer one of the following questions:

1. Explore, in your own words, the controls of one of these devices, and be at ease with them. What are the most important controls? Which ones will you use the most? Why?

2. Analog video sources need to be timed into a switcher so that we can do synchronous transitions between them. Waveform monitors and vectorscopes can be used to facilitate this timing. Explain in your own words how to do this.

CHAPTER 8
MONITORS AND TELEVISION SETS

250 words

Answer one of the following questions:

1. There are many new technologies out there to view television images. State some pros and cons of using each of them.

2. There's no getting around having to use television monitors, so we should know how to line them up—at least the essentials—what the controls are for, and how they affect the picture. In your own words, describe how to simply line up a colour television monitor, explaining why you're doing each step.

CHAPTER 9 CAMERAS

250 words

Answer one of the following questions:

1. Cameras use CCDs—their basic construction is what makes them unique and useful. There are variations on CCDs to give us single chip, two chip and three chip cameras. Explain in your own words how one of these works, and why it's important to know this.

2. Know your camera and its operational controls. Using one of the cameras you'll be using in your lab work, discuss the main controls you will use the most, explaining what to do with each one.

3. Cameras can be mounted almost anywhere. Where and how? What would be the best application for various types of camera mountings. Why?

CHAPTER 10 — LIGHTING

250 words

Answer one of the following questions:

1. Controllable lighting has been with us for some time. If you can control it, you should have a reliable means of measuring it. Explain how to use a lightmeter for basic lighting applications—give examples in your answer.

2. More sophisticated lighting fixtures give us more control—know your fixtures and what kinds of light they give out. State the three most common lighting fixtures you will probably use, and give examples of each of their uses. Explain why you would choose a certain fixture over another type.

3. Television has some limits—we can't just show absolutely anything that our eyes can see. What are these limits? How would you get around them? Give examples.

4. Changes in colour temperature are all around us, but since our eye adapts so well, we don't notice these . . . but television cameras do. Be aware of colour temperature and how we can work with it to achieve our creative results. Give examples of colour temperature issues and how you would overcome them.

CHAPTER 11

SPECIAL EFFECTS

250 words

Answer one of the following questions:

1. Switchers are the video equivalent of the audio console and associated effects equipment. Their complexity has evolved from wanting to cut between video sources, to dissolving, to wiping, to layering one source over another (keying). There are basic principles common to all of these effects. The better you are able to understand how it works, the more you will get out of your video program, regardless of what position you hold in the production. With that in mind, list the five most common switcher controls, and give examples of using them to create a visual effect. Explain why you chose the controls that you did.

2. DVEs have been around for about 25 years and have evolved, due to advances in technology and the wants and desires of production personnel. List five basic DVE effects and how you would express succinctly what you want without slurring your words.

3. Graphics generating devices (stillstores, character generators, paintboxes) have grown into multi-purpose, often PC-based, integrated systems. Describe three effects that you would want to create using the devices available to you in your lab, and explain how you would achieve them.

CHAPTER 12
VIDEO RECORDING AND REPRODUCING

250 words

Answer one of the following questions:

1. Videotape recording has been around for about 50 years. There have been various analog formats created over the years, and some of them are still with us today. Describe two legacy VTR formats that you might find still in use, and explain why they would exist in a contemporary production facility. Provide proof for your answer.

2. What is the latest technology in digital video recording? Why is this an important innovation? Provide sources for your answer.

CHAPTER 13: EDITING FOR TELEVISION

250 words

Answer one of the following questions:

1. Time code can be recorded in two different ways. State the differences between them, and explain the differences between drop frame and non-drop frame code, noting a real-world application of this knowledge as an example.

2. Educate yourself about the capabilities of the nonlinear editing software you use in your lab, and the aesthetic differences as well as innovative production approaches used in these suites. Describe five of the most frequently used elements of your nonlinear editing system. Explain why you chose these and why they are important.

CHAPTER 15 — TRANSMISSION

250 words

Answer one of the following questions:

1. For best results, antennas should be of a certain length and polarity. Explain why this is important to know, and give real world examples of applying this knowledge.

2. The electromagnetic spectrum has been carefully carved up by the powers that be so that everybody gets to use what they need and not interfere with one another. The dissection is intricate, and you should be familiar with some of the fundamental slices. List some of the basic segments of the electromagnetic spectrum that are used by professional broadcasters, in order of frequency, and state why this is important to know.

3. Master control is the place we've all been waiting for. Without it, all of our hard work never makes it to air. Be conscious of the importance of this facility and some of the things this environment does to keep us in the skies. List the main functions, indicating how the operation of MCR in these functions should normally go, and highlight potential ways in which they may go awry.